Hypothesis–Experiment Class (*Kasetsu*)

Hypothesis–Experiment Class (*Kasetsu*)

By

Kiyonobu Itakura

Edited by

Haruhiko Funahashi

A collaborative effort by the *Science and Method* Translation and Publication Committee

First published in Japanese by Kisetu-sha, Tokyo in 1969 as *Kagaku to hōhō* (『科学と方法』).

This English edition published in 2019 jointly by:

Kyoto University Press
69 Yoshida Konoe-cho
Sakyo-ku, Kyoto 606-8315 Japan
Telephone: +81-75-761-6182
Fax: +81-75-761-6190
Email: sales@kyoto-up.or.jp
Web: http://www.kyoto-up.or.jp

Trans Pacific Press
PO Box 164, Balwyn North
Victoria 3104, Australia
Telephone: +61-(0)3-9859-1112
Fax: +61-(0)3-8611-7989
Email: tpp.mail@gmail.com
Web: http://www.transpacificpress.com

Edited by Teresa Castelvetere.

Designed and set by Sarah Tuke, Melbourne, Australia.

Book cover designed by Takanori Hirano, Chiba, Japan.

Printed by Asia Printing Office Corporation, Nagano, Japan.

Distributors

Australia and New Zealand
James Bennett Pty Ltd
Locked Bag 537
Frenchs Forest NSW 2086
Australia
Telephone: +61-(0)2-8988-5000
Fax: +61-(0)2-8988-5031
Email: info@bennett.com.au
Web: www.bennett.com.au

USA and Canada
Independent Publishers Group (IPG)
814 N. Franklin Street
Chicago, IL 60610
USA
Telephone inquiries: +1-312-337-0747
Order placement: 800-888-4741
(domestic only)
Fax: +1-312-337-5985
Email: frontdesk@ipgbook.com
Web: http://www.ipgbook.com

Asia and the Pacific (Except Japan)
Kinokuniya Company Ltd.
Head office:
3-7-10 Shimomeguro
Meguro-ku
Tokyo 153-8504
Japan
Telephone: +81-(0)3-6910-0531
Fax: +81-(0)3-6420-1362
Email: bkimp@kinokuniya.co.jp
Web: www.kinokuniya.co.jp
Asia-Pacific office:
Kinokuniya Book Stores of Singapore Pte., Ltd.
391B Orchard Road #13-06/07/08
Ngee Ann City Tower B
Singapore 238874
Telephone: +65-6276-5558
Fax: +65-6276-5570
Email: SSO@kinokuniya.co.jp

ISBN 978–1–925608–87–8

The publication of this book was supported by a Grant-in-Aid for Publication of Scientific Research Results (Grant Number 18HP5257), provided by the Japan Society for the Promotion of Science, to whom we express our sincere appreciation.

Contents

Preface

This book represents the English translation of four seminal theses on the concept of 'Hypothesis–Experiment Class,' written by its originator and main proponent, Dr. Kiyonobu Itakura.

Hypothesis–Experiment Class (hereafter HEC) is, in its simplest terms, a class designed so that students can learn 'basic, universal scientific rules and concepts' in an enjoyable way. The class design is also continually and carefully cultivated in the hope that its results remain independent of incidental factors such as differences in students' backgrounds or in teachers' expertise. One final hallmark of this class is the HEC 'Classbook' – a combined textbook, reader, notebook and teacher's guide all-in-one, simultaneously fulfilling the needs of learners, teachers and researchers.

The particular learning process central to HEC is founded on research in the field of science history as well as on theories of cognition; it entails learners re-experiencing the investigative processes of prior, eminent scientists. Thus, students are given significant freedom to come up with their own ideas, debate, and change their mind for each problem; and 'leading' the students is completely eliminated from the teacher's duties, so that students come to judge what is correct or incorrect based on the experiment. Another major point of distinction between this class and conventional pedagogical theories is the fact that evaluation of the class, particularly the objective standard for success or failure of the class, is centred on the learner rather than the instructor.

In the 1980s, HEC was introduced in Western academic circles due to the efforts of Giyoo Hatano and Kayoko Inagaki, Japanese pioneers in the field of cognitive science. Articles by Hatano and Inagaki gained attention from researchers especially in the field of cognitive science and science education. In these fields, HEC was often known by the name 'HEI,' which stands for 'Hypothesis–Experiment Instruction.'

Hatano and Inagaki reported that the appealing structure of HEC, consisting of problems with several alternative responses, inherently helped students to gain a deep understanding of the world. The discussion stage of classes acted to socially amplify student motivation, and to provide continuous comprehension activities for the class. Through the problem-solving sequence of Expectation–Discussion–Experiment, students could grow their understanding. Many researchers were impressed by the pair's research focused on the productive role of social interaction in fostering deep understanding.

On the other hand, to date, previous researches have not focused sufficiently on the HEC Classbooks which contain problems carefully ordered to guide the understanding of scientific concepts underlying each problem set. In fact, the key to HEC strategy lies in the particular ordering of sequential problems. This approach, which is the guiding principle of this book, goes to the heart of Dr. Itakura's ideas.

We see in Dr. Itakura's theory a nascent appreciation of what is now an important keyword in science education – the idea of "conceptual change." The fact that 'children (all learners regardless of age) do not have zero knowledge or conception of scientific ideas, even ones that they have never been taught before' was demonstrated by Dr. Itakura in 1963, during his advocacy of HEC.

'Misconceptions' should not be simply corrected without due consideration; they have a crucially important role in improving the understanding and motivation of 'fellow learners (classmates).' Without sufficient appreciation of this, teachers may manage to find something that 'sparks an individual student's curiosity,' but will likely fail to cultivate the deep, persistent curiosity and social interactions so important in science.

This sort of premise leads us to a philosophical understanding that 'mistakes can only occur when the student has engaged in some type of thought' and that 'cognition is a social act.' By actualising that philosophical viewpoint, HEC has clearly demonstrated, throughout a large number of trial classrooms, that 'research in the field of science education is only possible through the discovery of *a series of well-crafted problems* and the development of a book containing those problems and related reading materials in proper arrangement — that is, the HEC Classbook.'

We should not focus primarily on 'interesting problems that motivate students.' Providing students with the HEC Classbooks which present well-crafted problems and readings is of vital importance for establishing true scientific recognition.

Through our research, we found that 'children genuinely enjoyed science and looked forward excitedly to having the class again.' That was indeed our aim from the beginning. However, we did not initially expect that it would also result in 'a lot of teachers discovering the joy of being a teacher.' This finding, which emerged through the trial classes, spurred us on to research the 'role and joys of being a teacher.'

The robust approval from learners and teachers alike derives from 'discovering the fun of science and its usefulness.' We believe that a wider appreciation of these kinds of results from HEC will lead to rapid developments in science education for all societies and classrooms across the globe, and it is with that conviction that we pursued the publication of this book.

The theses that constitute this book represent only a tiny portion of Dr. Itakura's corpus: a mere four chapters from *Science and Method*, one of the earliest of the over 200 books that he has written. However, we strongly hope that this book spurs the emergence of many revolutionary researchers in science education in the future.

Lastly, we would like to add that from 1963 to 2016, the natural science HEC Classbooks have expanded to include over fifty titles, primarily under the guidance of Dr. Itakura himself. Currently, there are not only natural science HEC Classbooks, but also those for *Money and Society, Prohibition and Democracy,* and *Flags of the World*, as well as many titles for the social sciences relating to subjects such as democracy, economic principles and world history. All Classbooks are tested and revised in as many trial classes as necessary to assure their quality. The HEC Classbook has continued to exert a significant influence on education, including in fields beyond the natural sciences.

Saburo Takeuchi, Spokesperson for the Association for Studies in Hypothesis–Experiment Class (ASHEC)
August 1, 2018

Acknowledgements

Dr. Itakura, the original author of these theses, had been quite pessimistic about translating his work from Japanese into any foreign language. He indicated his primary reason for this as follows: 'I want to devote all my time to my research into combining science history and science education in order to establish a new science, that being the science of science education.' He also jocularly offered that, 'if everyone in the world can read about Hypothesis–Experiment Class (HEC) without learning Japanese, then they have one less reason to *ever* learn Japanese!'

However, inquiries regarding HEC started arriving more and more from overseas. Even within Japan, increasing numbers of people were eager to put HEC into practice in other countries, so as to prove that 'anyone, anywhere, can do it and enjoy it.' In order to meet this demand, a few members of the Association for Studies in Hypothesis–Experiment Class independently began planning for the translation of Dr. Itakura's papers and HEC Classbooks in 2006. When we sought Dr. Itakura's permission, he gave it on the condition that we go about the translation 'hypothesising and testing.' We would again like to express our deep gratitude to him here.

From that point on, we provided isolated translations of lessons from the HEC Classbook to individuals upon request but met with slow progress regarding the official translation of Dr. Itakura's original theses, despite our enthusiasm. The translation was finally completed once we obtained a keenly sympathetic and knowledgeable translator, Alexander Clemmens. At the final stage of the translation work, Dr. Itakura himself also participated in discussions on the translation of several important terms from his bed and expressed his appreciation for the publication of this book. We also want to thank the bereaved families who continued to give unfaltering support to our committee after Dr. Itakura passed away.

We would like to express our gratitude to Tetsuya Suzuki of Kyoto University Press in Japan, and Masaaki Ogawa of the Kyoto University

Research Administration Office for turning our manuscript into a book. We do not have enough words to thank Yoshio Sugimoto and Teresa Castelvetere of Trans Pacific Press in Australia for their patience in helping us as we struggled with editing in English, which is not our native language. We also wish to thank Nobuo Takahashi for his assistance as an English to Japanese translator throughout the translation and review process.

Lastly, we have received considerable help from an unexpectedly large number of people, whether in the form of supporting us with their wisdom or fundraising efforts. We are unable to mention each individual by name here, but we certainly wish to express our heartfelt thanks to each and every one of you. Thank you.

Science and Method Translation and Publication Committee
Haruhiko Funahashi
Saburo Takeuchi
Mariko Kobayashi
and Alexander Clemmens

Biography of Dr. Kiyonobu Itakura

1930 Born in the old downtown area Tokyo.

1953 Graduated from the Department of History and Philosophy of Science in the College of Arts and Sciences at the University of Tokyo.

1958 Completed studies (Physics) at the Graduate School of Mathematical and Physical Science at the University of Tokyo. Obtained a PhD in Physical Science, specialising in Science History.

1959–1995 Entered the National Institute for Educational Policy Research, where he was the chief of the Physics Education Research Laboratory.

1963 Proposed 'Hypothesis–Experiment Class.' Inaugurated the *Kasetsu Jikken Jugyō Kenkyūkai* (Association for Studies in Hypothesis–Experiment Class). Wrote a number of 'Hypothesis–Experiment Classbooks' with the help of many educators conducting trial classes.

1995 Retired. After retirement, Dr. Itakura established the private 'Itakura Research Center,' where he conducted science education research in a wide range of fields, while publishing many works. He was also the chief editor of '*Tanoshii Jugyō*' Magazine from 1983 to 2018.

2013–2017 Chairman of the History of Science Society of Japan.

2018 Dr. Itakura died of old age in Tokyo, on February 7.

About the editorial committee members

Professor Haruhiko Funahashi

Professor Funahashi earned a doctorate in Science from Kyoto University in 1997. Currently, he works as a professor at the Institute for Liberal Arts and Sciences, Kyoto University. In the field of fundamental physics, he has conducted experimental studies of elementary particles and the universe.

In his course in the general education curriculum, his focus is on students who do not major in science having a chance to cultivate scientific, critical minds through learning physics. In order to achieve this, he adopts the concept and methods of *Kasetsu* (HEC) in his interactive class titled 'Physics for all.' He has led the translation and publication project of the present volume from 2016.

Theses

Interferometer for cold neutrons using multilayer mirrors. *Phys. Rev. A 54,* 649, 1996.

Survey of Japanese High School and University Students using the Force Concept Inventory Translated into Japanese, *Proceedings of the International Conference of Physics Education 2006 – Toward Development of Physics for All-(ICPE2006),* 2008.

'I have found the pleasure of repeating predictions and experiments': A practical report of 'Physics for All' with Hypothesis–Experiment Classbooks, *Tanoshii Jugyō,* (No.366), pp.107–116, 2010 (in Japanese).

Saburo Takeuchi

Saburo Takeuchi graduated from the Japanese Literature Department of Nihon University in 1963 and worked for a publishing company, where he met Dr. Itakura. Since then, utilising his talent as an editor, he has been a staunch supporter of Dr. Itakura, remaining with him throughout the development and establishment of HEC theory. In 1973, he founded a publishing company,

Kasetu-sha, in Tokyo, which has worked on the publication of books related to HEC and Dr. Itakura's works.

After Dr. Itakura passed away, since August 2018, he has been the spokesperson for the Association for Studies in Hypothesis–Experiment Class.

Mariko Kobayashi

Mariko Kobayashi majored in Agricultural Science and graduated from Meiji University in 1974. She learned HEC in the first year of her career as a junior high school science teacher and has been putting it into practice since 1987. Her works have helped children broaden their interest in science in a wide range of fields. She was also a collaborator of Dr. Itakura in his designing of the Classbook '*Limestone, the Mysterious Stone.*'

After participating in the International Conference on Physics Education (ICPE) 2006, she proposed the organisation of an English translation project team for Dr. Itakura's thesis and HEC Classbooks. Since then, she has managed the secretariat of the project.

Works (in Japanese)

Experiment Guide for 'Limestone, the Mysterious Stone'. Kasetu-sha, 2009.
Anatomy Classroom for Niboshi (boiled and dried fish). Kasetu-sha, 2010.
Have you ever Seen Cosmic Rays? Kasetu-sha, 2018. (Joint work with Miyuki Yamamoto)
If you Could See Atoms, with computer simulations. Kasetu-sha, 2008. (Joint work with Dr. Itakura)

Alexander M. Clemmens

Alexander Clemmens studied chemistry at Carnegie Mellon University, Pittsburgh, USA. He graduated from the university in 2006 and majored in inorganic chemistry at the graduate school of the University of Pittsburgh (2008). He was accepted into the JET program (Japan Exchange and Teaching Programme), and came to Japan in 2009, working as an ALT (Assistant Language Teacher) for two years.

Currently, he works as a translator with several companies, mainly in the fields of medicine, engineering, science and science education. After doing several English reviews of translations of HEC Classbooks, he joined the editorial committee at the request of the project team and Dr. Itakura. Beyond his role as simply a 'translator,' Alexander Clemmens has engaged in lengthy discussions about many of the finer points of the text and the corresponding translation, thereby deepening our own understanding of HEC.

Certificate
Japanese Language Proficiency N1
The Japan Kanji Aptitude Test Grade 2

A Collection of Articles and Essays by Kiyonobu Itakura

1 The Process of Establishing Mental Recognition in Science

1.1 Conditions for establishing mental recognition

Recognition[*] (of a physical object or scientific truth) is established only through testing and experiment.

Here we use the terms 'testing' and 'experiment' in general to refer to human activities that act on an object or event with investigative intent, and 'experiment' specifically to refer to any activity that aims for recognition of the object or event itself.

The above proposition is our attempt to distinguish *recognition* from mere 'perception.' When the retina produces an image from light bouncing off an object, we could call this 'perception' but not 'recognition.' True mental *recognition* of an object requires the perceiver to conduct some action with intent to confirm something about the object.

For example, many people gaze up at the Moon but probably do not consider or wonder about how it waxes and wanes. They may stare at the Moon and perceive its shape, but if they do not try to

---

* Translator's Note: The idea of 'recognition' central to the author's thinking does not connote the meaning 'praise from or acceptance by peers' but is epistemological in nature and refers to the cognitive process of perceiving something and knowing what that something is. We have occasionally translated it as 'scientific knowledge' when the context allows for it to clearly indicate the epistemological nature of 'knowing' and there is no danger of the phrase being interpreted to mean 'current scientific facts.'

investigate anything in particular about that shape, their action does not result in 'recognition' of the Moon. To put it simply, *recognition* can be defined as investigated perception. We would even go so far as to say that recognition is established the instant an individual's perception includes any element of investigation (expectation).

Next, 'action' with investigative intent performed on an object, i.e. an experiment, does not necessarily mean physical interaction with the object.

The term 'action' here includes activities like prediction and inquiry, even seemingly trivial questions like, 'Does that phenomenon *really* happen?' Confirming the accuracy of a prediction and re-examining the object can also be considered types of 'action.' Therefore, the scientific activity called *observation* can be referred to as an experiment. However, not all types of 'observation' constitute an experiment as we have defined the term here. For example, merely making careful observations of an object (such as cherry blossoms) without any investigative intent is excluded from our definition of experiment. Conversely, any act of observation which includes a concrete prediction or investigative intent, e.g. 'I wonder how many stamens a cherry blossom has,' is included in our definition of *experiment.*

In summary, the term *experiment* as used here includes any activity normally designated as 'observation,' such as observing plants or stars, as long as that observation is performed as an investigation. This definition centers the idea of *experiment* around the human process of recognition.

Hands-on trials are not usually considered experiments because they are not performed for the purpose of recognition. However, since the person performing any trial will ultimately have to check the results against their initial expectations or bias, the trial may be treated as a type of *experiment.*

In the past, educators have often claimed that making predictions before an experiment should be forbidden, because it creates a bias in the experiment and may hinder disinterested observation of facts. But, we must state here that this is a mistaken belief, one which arises from insufficient understanding of the process of mental recognition in science. Experiments without an active will to understand something are not true experiments. People are free to ignore any natural phenomenon, even one that happens right in front of them as, indeed, often occurs during student experiments. However, a 'prediction' remains as mere 'bias' unless there is an experiment at some time to test it.

Prediction and bias could be interpreted as having nearly the same meaning, but most people would call *bias* a belief which need not be confirmed by facts. Therefore, we shall distinguish the word 'prediction' from 'bias.' *Prediction* is belief recognised as requiring confirmation, and *bias* is a type of dogmatic belief held without concern for the necessity of confirmation. Using this definition, it is impossible to perform an experiment with a 'bias' regarding the result. Furthermore, this definition also clarifies that what establishing recognition requires is a *prediction*, not bias. Even so, we cannot claim that bias is completely useless for developing recognition. A person clearly aware of their own bias, if faced with an irrefutable fact that contradicts that bias, might be able to quickly and cleanly reject their bias and recognise the truth. In addition, since the distinction between bias and prediction is somewhat blurry in the first place, they cannot always be clearly separated. Frequently, one's bias becomes a good target for investigation (prediction) once opposing evidence crops up. Thus, having a bias will not significantly hinder establishment of mental recognition; rather, not having such a bias or prediction will. As Heraclitus, a philosopher of Ancient Greece, stated, 'He who does not expect will not find out the unexpected, for it is trackless and unexplored' [Charles H. Kahn, *The Art and Thought of Heraclitus*].

Certainly, predictions and biases may also negatively influence a person's immediate perception of facts, or stymie later recall of experimental results, but the present discussion is not concerned with these problems.

1.2 Hypotheses and mental recognition in science

Recognition in science is a recognition of rules or principles, meaning that it should be a recognition of universality, which allows us to make predictions about unknown phenomena.

In principle, the accuracy of scientific propositions must be tested by some objective means at some point. Recognition in the scientific sense must aim to make the law-like behaviour of the phenomenon clear; that is, the same conditions always yield the same phenomenon. We should not focus on individual events which occurred only once and cannot be duplicated using any principle. Potential scientific targets for recognition include the history of the Earth, the origin of life, the evolution of living things and natural history. These fields may start out as merely descriptive history, propped up by the research methods and results of other sciences, but eventually they must outgrow simple description and discover general principles at work in the Earth's evolution. How life was able to arise on the Earth, despite the remoteness of this event, is an issue that ought to be investigated in terms of the conditions that prevailed in that bygone era. Such investigations should always proceed with the tacit assumption that given the same conditions, history would follow the same path. That is why experimental research remains paramount even in the research of historical science.

Law-like recognition can only be achieved through experimental attempts to prove a hypothesis. This statement means that simple explanations of several well-known phenomena do not constitute *science*; an explanation must be offered as a hypothesis and then proven by experiment to become a scientific law or principle.

Scientific propositions aim for law-like recognition of an object and not merely isolated knowledge about individual events or facts. Therefore, scientific propositions include not only stipulations of known facts but also predictions and precepts for a range of future or unknown events. Therefore, it is possible to check the veracity of these propositions by comparing the predictions based on these propositions with the facts of those actual events.

Scientific theories, laws and hypotheses must contain specific and unambiguous claims which can be tested through experiment.

In prior research on science classrooms, the so-called theories or hypotheses offered in class did not contain measurable or unambiguous claims, such that verifying these claims through experiment was impossibly challenging. Scientific laws and theories only have existential value when they make specific predictions about some unknown event. Similarly, hypotheses only have existential value when they lead to that type of law or theory.

Once a hypothesis and its concurrent theoretical prediction are found consistent with a range of experimental facts, the truth of the hypothesis can be affirmed. It then attains the label of theory or law.

Scientific propositions are based on a limited number of known events and make stipulations and predictions about an unlimited number of unknown events, so we can never be absolutely sure in principle that the described scientific propositions will always hold true. Yet, the history of science teaches us that a theory or principle demonstrated in a small number of carefully scrutinised experiments often proves to be a theory or principle with extremely wide applicability and usefulness. However, the history of science has also shown that uncritically accepting a theory or principle seen in a small number of experiments may lead to gross error, and that it is impossible to determine how useful any theory will be ahead of time. Newton's laws of dynamics provide a good example. Having been established through a relatively small number of experiments, these laws then proved to be extremely useful as the foundation for a wide range of theoretical and experimental research thereafter. Still, their limitations were revealed with the advent of the theory of relativity and quantum mechanics. We must remember that theories are true only within applicable limits. This is fundamentally important for science education.

There are two types of hypotheses. Some hypotheses are difficult or time-consuming to prove, so they are provisionally accepted as true while gradually being developed into a theory. Other hypotheses may be easy to test directly, but – regardless of whether they are true or not – offer a useful conduit for further inquiry, so they are adopted as temporary or utilitarian guides. Some prefer to call the latter type a 'working hypothesis' and only refer to the former as a genuine hypothesis, but both are included in our use of *hypothesis* unless otherwise indicated.

A *working hypothesis* is a proposition which has already been experimentally verified or disproved through research, but which is accepted as a temporary and utilitarian tool allowing researchers to test it on actual objects. These working hypotheses are not widely known to the public, so the word "hypothesis" is often seen as referring to hypotheses like Avogadro's Hypothesis or the wave-particle hypothesis for light, which do not have years of experimental evidence behind them but are taken as true to build a theory and spur further research. However, our view is that hypotheses in general play a defining role in recognition, so we deliberately include working hypotheses in our definition of *hypothesis*.

People may have trouble distinguishing a *hypothesis* from an *expectation* (or prediction) so we offer the following explanation. A *hypothesis* is a prediction about a more-or-less general theory or principle, whereas an *expectation* is a prediction about an individual unknown event. If you say, 'I think it'll be sunny tomorrow' as an isolated prediction, you have an expectation but no hypothesis. But if you were to say, 'I think Theory X is true; therefore, Event Y will happen,' then you would have both a hypothesis and an expectation. If a series of expectations based on this hypothesis proves true, the hypothesis is verified.

Now, it is true that Newton famously said, 'I don't contrive hypotheses' (*hypotheses non fingo*), but he wrote this with the Cartesian framework in mind, in which hypothesis means 'nonsense that cannot be proven through experiment.' Newton himself performed careful research on the photon hypothesis of light and other hypotheses.

Through the process of verifying predictions about the natural world, hypotheses allow us to discover aspects of nature that are completely hidden or utterly confounding if we only look at the superficial, natural events we casually encounter.

The part of the natural world that we humans come into superficial, incidental contact with is often extremely limited, just one side of a multi-faceted reality. Other aspects of the natural world can only be discovered through a deliberate effort. For example, based on our everyday experience, the Earth is flat and not round. The Sun seems to be revolving around us (the Earth), and the Earth does not seem to be moving at all. All of this is because humans live on one tiny part of the big Earth and cannot observe it from the outside. Yet, by establishing various expectations and hypotheses about how the heavenly bodies might be moving, and comparing those expectations with the facts, we were able to discover that the Earth is in fact round and that it rotates around the Sun, even during an era when we could not directly observe those realities. This is how the establishment of hypotheses allows us to make correct judgments which surpass the findings from a single incidental facet of nature.

Often, it is not possible to determine unequivocally how to properly interpret a series of facts or what is the best theory to develop from them. Further experimental proof is usually necessary to determine which interpretation is correct.

In general, finding an unambiguous, unlimited, true theory from a limited set of facts is no simple task. There are likely to be multiple possible interpretations available regardless of however many facts one has assembled; thus, experimental proof becomes necessary. Science does not progress through a process of *fact → theory*, as is commonly believed, but through a cyclical process of *hypothesis → experiment (facts) → hypothesis → experiment (facts).*

The same can be said in science education. Conventional science education practice has been to first show a series of experimental facts, and then to inductively reason out a theory based on these facts. The inductive education method has been seen as ideal and used extensively; and yet, however straightforward the experimental results may seem, it is often not possible to determine a single best interpretation. At least, students will be likely to feel this way.

Even students with minimal knowledge of the topic at hand will nonetheless probably have, at least, some naive ideas regarding the topic. These may not be very logical and may even be an amorphous understanding with fluid interpretations. Despite being presented with a new *experimental* fact, the student will often spin out a newer interpretation based on their everyday understanding just to incorporate the new fact. And yet the teacher and textbook pay no heed, simply concluding, 'So we can see that these facts lead to such and such.' In cases such as this, the teacher has brutally forced a theory on the student using the experimental facts as a shield. Naturally, the students are left with a defiant or sceptical attitude. In comparison, it would be less objectionable to push the theory by stating, 'This is what seems to be true based on research performed by scientists in the field, so let's just learn it for now.' This does not crush the students' will as much and is therefore less damaging.

Experiments are conscious, agent-based investigations of the natural world which require joint control of self and nature and are for the purpose of mental recognition of some aspect of the natural world.

We touched on this at the beginning, but we think it is important to re-emphasise here the meaning of hypothesis and experiment (as well as their relationship with each other) in science and science education. Conventional education has the wrong idea of 'experiment,' and this has caused a great deal of unnecessary confusion.

The above proposition is our attempt to clarify that simply using experimental equipment to replicate some natural phenomenon is *not* an experiment, despite what people generally imagine. If you do not have agency or investigative intent, no amount of fiddling with apparatus counts as an experiment. Give chimpanzees or toddlers some experimental equipment and they will play with it, but they are clearly not conducting any experiment. Nonetheless, this is exactly what we see in the 'experiments' performed in traditional science classrooms. Many students had no idea why they were using that particular equipment for that 'experiment' or what the 'experiment' was examining. Regardless of whether the 'experiment' was performed by the teacher or students, these students just stared at the experimental operation, occasionally moving the apparatus as ordered.

To make *experiments* worthy of the name in science education, we must ensure that every student maintains investigative intent, clearly cognizant of the yes–no questions they are asking in the experiment. Thus, before engaging in the experiment, every student must be made to have an expectation (hypothesis) that they bring to the experiment.

The key criterion for stating that a person has sufficient understanding of a principle at work between certain concepts is that the person can use the concepts and principle to make predictions about unknown events within the scope of the principle.

The theories and principles of science are expressed in relatively concise terms, so it is not too difficult to simply memorise them. However, those theories and principles are distilled from a wide array of relevant research, and it is not necessarily easy to understand the research itself. Therefore, science education should teach students about the specific facts that a theory describes and also how much predictive power the theory has. Our goal then is to get every student making correct predictions using a concept or rule by taking them through cycles of *expectation* → *experiment* → *expectation* → *experiment* for each problem that is set. If we cannot achieve this, we cannot say that we have taught all the students in class the target concept or rule.

1.3 Science as social recognition

Recognition in science is a social recognition.

Science is not something that can exist solely within an individual human; it exists as a part of society. In light of that understanding, we cannot talk about scientific knowledge in the individual as separate from society. As we elaborate below, this also means that mental recognition as a part of science is only established when it is social in nature.

Firstly, if someone discovers a principle but does not publish it and have it accepted by society, we cannot say that the principle is a part of science.

'Be accepted by society' can include various levels and need not mean one specific thing. It is impossible to have all members of society understand and affirm every single natural law or theory. Instead, there are organisations (academia and scholarly groups) who are entrusted with managing scientific knowledge and who can function as a means of preparing such scientific knowledge for social consumption. They are bodies with authority, so we can effectively say that 'be accepted by society' means 'being accepted by a body of scientists.' Certainly, it is rare for every scientist in an academic group to immediately accept the validity of new scientific knowledge. A report or paper by an individual is first registered as material to be reviewed by peer scientists. This is the point when the knowledge (under peer review) first becomes a part of science.

The first requisite for 'being accepted by society' is that a research report be published and made available for critical examination by others. In the history of science, there are plenty of cases in which a person was ahead of their time and discovered a valid law or scientific theory, but – due to a lack of confidence, fear of backlash, or a lack of interest in contributing to society – they did not make their finding public. No matter how extensively any such theory was proven in research, it cannot be considered a part of science until it is published in society.

On the other hand, there have been scientists who did not happen to know about a well-established theory or principle, and thus inadvertently duplicated that research, rediscovering the same principle. The researcher may imagine their work to be unique, but unless a novel method was employed, their efforts must be considered to be of no value to science, because the researcher will have added nothing new to science in society.

The job of a scientist is not to talk about truth; but, rather, to demonstrate a truth in such a way that anyone could acknowledge it.

However much confidence a researcher has in the validity of their discovery, if it has not been sufficiently proven in a form acceptable to other scientists, the research is still inadequate. Furthermore, even if all people presently on Earth accept the conclusion as true, but there is no means of proof sufficient for any hypothetical individual in society to accept it, it is not yet science.

Therefore, we cannot claim that a theory or law which has not been demonstrated in a form sufficiently acceptable to a student in a classroom is a part of science for that student.

Anyone wishing to advance scientific thinking may freely borrow socially accepted scientific knowledge and must build any new theories with that stockpile of scientific knowledge as a foundation.

Scientific research does not progress when a researcher tries to investigate a matter entirely by themselves, but rather when they first read the research papers of people who have already examined that topic and then build upon their expertise. Science is not some vain duel of knowledge; it exists to increase knowledge in society. If your results differ from those of another researcher, posing the issue to see which results are right is indispensable for making the science into something social. We should not refrain from critiquing or criticising the research of others. Conventional science education has not done a good, conscientious job of describing the importance of borrowing the wisdom of others and sharing your own. It is, therefore, even more important to train students in how to benefit from the wisdom of others than in how to make discoveries on their own.

A theory, even a well-established one, is just a hypothetical idea when a person first tries to understand it.

Science is social and has been organised by groups of scientists into scientific propositions accepted by society as valid and useful. Members of society can trust those scientific propositions and make free use of them. That means that people should be able to adopt those rules easily even when first coming into contact with them. The fact that even school children today can easily grasp heliocentric theory and the existence of atoms or molecules (which depend on abstract scientific systems) is because these ideas have been accepted in society.

Simply adopting a socially-accepted theory and actually understanding that theory are different, though closely related, issues. Many people who genuinely believe they have accepted a given scientific theory may still adhere to an understanding clearly contradictory to the theory. Additionally, common sense and myths also have a type of unique social authority, so whenever any of these rationales come into conflict, social authority loses some of its influence, and the individual must investigate the matter for themselves.

In fact, the early stages of science education often present scientific theory and knowledge that is new to students and contradicts their individual preconceptions or common sense (although this has not been widely recognised in conventional science education). In these cases, regardless of how well established the theory or law is in science, it should be presented as simply another hypothesis, one equivalent to the students' preconceptions and common sense. In this way, the students will be able to experience for themselves the process of arriving at an understanding that the scientific theory is a far more accurate and useful way of examining the issue.

This point was largely overlooked in the past, probably bringing about many of the failures in conventional science education. The preconceptions and common sense present in preschoolers' minds were ignored. If the scientific theory or law was not presented in a way that clearly revealed the conflict with the student's preconceptions, the student would end up with both conflicting ideas present in their head. Students would then keep the commonsense thinking for normal life and use the knowledge learned in science class for school tests. Some did this without realising that these ways of thinking were contradictory, but many did this fully aware of the contradiction. They did this because they would just get confused and had to separate the conflicting rationales into different spheres of life. Worse, students' preconceptions have taken root through various experiences in life and are likely to be more plausible to them than some scientific theory learned in an experiment at school.

New scientific theories and discoveries often reject or contradict common sense, preconceptions or widely accepted science; therefore, new theories must be presented with the necessary readiness and proper support, in the form of a clear basis that allows any objection or doubts to be overcome. Making a new scientific discovery, i.e. something that brings new knowledge to society, means bringing knowledge that is unexpected or even unwelcome to society's knowledge base. Therefore, researchers need the wherewithal to question the accepted wisdom of science and, occasionally, to push past society's objections if they wish to do truly original research.

The more that scientific research and its findings overturn widely accepted science, common sense or preconceptions, the more that research is regarded as novel. A position in society in which one is free to flout preconceptions and a will strong enough to overcome objections are prerequisites for this kind of groundbreaking research. Thus, science education must train students to handle these kinds of social aspects in addition to training their ability to research the natural world.

2 What is the Hypothesis–Experiment Class? History and Classroom Management

2.1 About the name HEC

Hypothesis–Experiment Class (*Kasetsu–Jikken Jugyō* in Japanese, hereafter HEC) refers to the science education materials and method based on the theory proposed by the author in 'The Process of Establishing Mental Recognition in Science.'

This type of class is called *Hypothesis–Experiment* because it fully embodies the philosophy previously introduced that hypotheses and experiments are the foundation of scientific knowledge.

It is further called a 'class' and not a 'program' because scientific knowledge is social in nature, as stated in 'The Process of Establishing Mental Recognition in Science.' That is our foundation for crafting the Hypothesis–Experiment concept as a group-based 'class' for science education in school and not as a 'program' for an individual learning science.

Lastly, HEC classes are not simply a classroom 'method,' because they require specific content about which students can hypothesise and subsequently investigate in experiments in order to discover some general theory for themselves. Furthermore, this type of class is made possible by the *HEC Classbook*

(*Jugyōsho* in Japanese), which supplies the content necessary for the educational goals of the science class.

The HEC approach attempts to cover the general basic concepts, laws and theories in a science curriculum as well as providing all the means for developing recognition in science. Teaching the most fundamental concepts, laws and theories does not, however, constitute the entirety of a core science education.

In order to make productive use of science in society, students will need more than the most general fundamental concepts, laws and theories; they must learn how to gain further specialised knowledge relevant to their particular field on their own. Specifically, this can mean developing an intuitive technical grasp of some special law or principle at work in a complicated situation. This requires ever-ready experience, confidence and ability. In that regard, educators should give thought to how other classes and HEC classes will fit together in their science curriculum. But we must leave that discussion for another time and place; here, we are simply clarifying the limitations of HEC.

2.2 Goals when implementing HEC

We have set the following three goals to be pursued in the implementation of HEC classes.

Goal 1: Make sure each and every student gains the ability to use the central theory or concept.

We have arranged the HEC Classbook so that, by going through the class process, all students (with special exceptions) should be able to correctly answer the final problem in any series of problems. Also, the class average on any cumulative test should be 90 per cent or higher, after careless mistakes have been revised.

The HEC process is (as described later) a sequence of *Problem → Expectation (Hypothesising) → Discussion → Experiment* cycles. The plan was to arrange problems in the HEC Classbook so that by the final problem in a unit, every member of class would be able to make use of the target concept or law to come up with the right answer before the experiment. Therefore, we had to create a trial version of the Classbook, conduct a large number of trial classes and revise it along the way. Our revisions were, of course, based on the idea that using a common Classbook would reproduce the same processes in any class in any school. With regard to our goal of 90 per cent, or higher, class averages on cumulative tests, the test problems must be completely within the scope of the target concept or law investigated in class. In the past, teachers have complained about poor performance on certain questions (in the section titled 'apply the principle') which actually went beyond the scope of the original target concept and required the students to make leaps. Teachers must diligently avoid these kinds of questions which go beyond the scope of the target concept. As a rule, the cumulative test should be given about one to two weeks after the final class. Students are apt to forget topics or mix up ideas after a unit is over, so educators may feel the need to give review problems as in a standard maths class. This is fine, as long as these review problems only cover what students have already learned.

Goal 2: Structure the class so that most students report that they like both science and these science classes.

The goal is to get over 50 per cent of students to freely offer that they 'liked' or 'really liked' science and the HEC and none of the students to say they 'disliked' it (with a couple of exceptions) in an after-class survey.

It is meaningless to get all the students to acquire the target concept and law for Goal 1 if they lose interest in class or learn to hate science. We should not sacrifice student interest by just repeating information over and over until slower learners understand everything. That is why we established this second goal: Goal 2 simply exists so HEC will not sacrifice any students on its way to achieving Goal 1. It must be a class enjoyed by all students. However, achievement of this goal is determined by students' subjective experience of the class, so it can be affected in complicated ways, including by their relative performance in the class or other, unrelated issues. Since delicate emotions are a factor with this type of evaluation, we have kept the benchmark down to merely the majority (over half) of the students reporting that they 'liked' or 'really liked' the class. Ideally, we want to aim for 80–90 per cent of students reporting that they enjoyed class, but given the difficulty of accurately assessing emotions, we decided to establish the 50 per cent benchmark as a minimum standard. On the other hand, there may be two or three children with special circumstances – lingering embarrassment from something that happened a few hours before, having been absent and not understanding the class, or upset that the class does not allow them to prepare ahead of time. We hope that no student, except for under these sorts of circumstances, reports that they 'dislike' the class.

Goal 3: Make all necessary preparations so that any teacher sufficiently passionate about education, not just special veteran teachers, will be able to teach this type of class.

We have structured everything so that the class does not require a great amount of learning or effort on the part of the teacher. Achievement of this goal will be judged on the basis of teachers' after-class feedback.

However effective a class may be from an educational standpoint, if too much effort or expertise is asked of the teacher, then the class results can be considered worthless from a research point of view. This is why we have established Goal 3.

What we mean by 'all necessary preparations' is that we will provide the scrupulously compiled HEC Classbook, as well as all test problems, enrichment problems, practice problems, readings, experimental tools, and reference materials for the class. The teacher does not need to make any special plans or prepare any reference materials. We conducted trial classes solely with the HEC Classbook and revised it numerous times. Based on the positive results from a large number of classes, we have compiled a HEC Classbook and class plan to our general satisfaction. This is why we feel confident that we can provide a more effective class plan – the HEC Classbook – than a class plan created by new or relatively inexperienced teachers.

Also, the requirement that the teacher be 'sufficiently passionate' is only present because, regardless of how well-crafted the Hypothesis–Experiment plan is, a teacher lacking the desire to understand and execute it will not be effective. Especially in the early stages, a teacher may be sceptical of this 'new-fangled' class and misunderstand its purpose, ultimately leading to a confusing or jumbled class. We felt a 'passionate' teacher would be necessary to avoid this. However, we have certainly made every effort to make these classes easily understandable to any teacher who wishes to use them.

The three goals above were established when the initial stages of research into these HEC classes were begun in September 1963. The three topics, *Motion of a Pendulum*, *Objects and their Weight*, and *Springs and Force* were then compiled into the HEC Classbooks and, as a result of over 150 classes with over 20 teachers and 2,000 hours of class, we can confirm that all three goals are achievable. So, at least for the aforementioned three topics currently in the HEC Classbook, we found clear proof that all classes were able to achieve all three goals.

We will leave detailed reporting of the trial classroom results for the time when we publish specific teaching materials.

2.3 HEC and the HEC Classbook

HEC classes centre around the HEC Classbook. For the sake of comparison, we can classify the textbooks employed in conventional science classes into the following categories based on form and function.

Reader-style textbook: The teacher has students read a section or passage, and then offers demonstrations, illustrations, questions or answers, to help the students understand the content and meaning of the passage. (Around 1886, several kinds of reader-style textbooks were published and in use in Japan.)

Summary-style: Information to be covered in class is provided in this kind of textbook. They are mainly used at the end of class as a summary and are also designed so that pupils can review the class by themselves. The archetype of these textbooks is the *Jinjō-Shōgaku-Rika-sho* [elementary school science book*], which was compiled by the Japanese Ministry of Education from 1911 to 1941.

Workbook-style: The teacher gives the students a task to complete, which they do on their own using the workbook. Afterwards, the teacher checks their answers. The workbooks have no printed answers or new information to learn. One typical example of this kind is the *Rika-Gakushū-chō* [science workbook] published after World War I, which was also used as a general notebook. Programmed learning worksheets are another typical example. Lastly, the *Shōtō-ka-Rika* [elementary level science], compiled by the Ministry of Education and used during World War II, was also this type of workbook, but shared some similarities with the HEC Classbook.

Pattern book-style: These are designed so that students are able to 'research' subjects for preparation or review, independently and without teacher explanations. The series of textbooks compiled by the Ministry of Education, *Shōgakusei-no-Kagaku* [science for elementary school students], first printed in colour after World War II, are typical pattern book-style textbooks.

The HEC Classbook, used in HEC classes, is unique among the textbooks listed above. It includes tasks for teachers as well as for students and guides the class with specific instructions; it is indispensable for HEC Classes. Its characteristics are quite

---

* Translator's Note: This and subsequent items in square brackets are literal translations added by the translator.

different from the conventional formats listed above, so we call it the '*Hypothesis–Experiment Classbook* (*Jugyōsho*)' in order to distinguish its function from that of the other books.

In conventional classes, each teacher designs their class based on experience or intuition whereas the HEC Classbook is designed to ensure reproducibility of class organisation. This reproducibility opens up the potential for systematic research and development of the class. The HEC Classbook was thus created to virtually guarantee classroom results that surpass the limitations of any individual teacher's skill.

A 'class' is basically just the interaction between the teacher of the class and its students, but there are scientific principles, independent of the teacher's personality or classroom dynamics, that are necessary to create the *best* class for enabling all of its students to grasp a fundamental concept, law, or theory in science. Those objective principles have been incorporated into the HEC Classbook which acts as a practical prescription for creating successful classes as efficiently as possible, regardless of teacher or classroom. Almost all classes run with the Classbooks developed to date have demonstrated good reproducibility of class performance, independent of teachers and students. All teachers of the trial classes have reported that the HEC Classbook reduces the mental and physical load of classroom management and helps them to create effective classes far more consistently than ever before.

2.4 How the HEC Classbook came to be

The *Problems* section is the focal point of the HEC Classbook, with *Enrichment Problems*, *Practice Problems*, *Questions*, *New Science Words (Definitions of Words)*, *Summaries of Scientific*

Laws, *Readings*, and *Materials* acting as secondary elements. Each of these elements which make up the HEC Classbook has its own respective role as follows.

Problems: A *Problem* is normally a four-stage process of 'the problem stated,' 'coming up with an expectation,' 'discussion,' and finally 'experiment.'** This process is intended to get all students making their own predictions, discussing their own ideas and then checking their ideas through experiment. *Problems* are arranged into this four-stage process on the basis of the author's theory of how recognition is established in science (see previous chapter).

This structure for *Problems* means that inquiries which cannot be clearly answered by experiment are not acceptable as *Problems*. The answer to a *Problem* cannot be decided by teacher authority or by class vote; it must be determined by objective experiment. However, this does not exclude problems that cannot be examined in easy classroom experiments. If predictions for a problem can be tested objectively with video or statistical materials, then it is acceptable as a *Problem*.

Also, these *Problems* do not require that students have any understanding of the relevant facts and principles beforehand. That is, there is no issue whatsoever with students making their first predictions based entirely on common sense, wild guessing, or preconceptions. The *Problems* are arranged so that participating in a series of problems on the topic will

---

** Translator's Note: The word 'problem' (in quotation marks in the original) referred to the whole four-stage process as well as to the first stage of the process in which the problem is stated. The 'making a prediction' stage is also referred to as 'expectation' or 'prediction.'

lead all students to independently make the correct prediction (expectation or choice) by the final stages. That is sufficient.

Questions: *Questions* do not require students to make any predictions to be tested in an experiment; their purpose is to inquire about a student's past experiences and memories. They do not require an answer from each and every student. As long as someone offers a reply, that is sufficient.

Questions are intended to introduce the class content by connecting it with students' prior experiences in the world. Any student who happens to have significant experience with the topic can simply describe it to the entire class. Also, when students do not have the theoretical experimental background necessary to understand something sufficiently, teachers may try to expand their horizons and pique their interest by asking for their wildest guesses before explaining what scientists have concluded. *Questions* can also include this type of inquiry. In conclusion, *Questions* differ from *Problems* in that *Questions* do not ever require a 'right' answer from the students.

Expectation: In most cases, the students are provided with *multiple choices* from which to choose their *Expectation.* This is to make it clear what type of prediction the problem is seeking, since it is not always clear just from the problem as presented.

Some educators may think that students should come up with predictions on their own, rather than be given ready-made choices. However, since students who have a fair grasp of the problem will come up with the same type of predictions as those ultimately intended, in the interest of time, we provide those choices in advance. Other educators may feel that choices could set up a leading-question type of situation, making the problem too easy, but that is not necessarily true. Similarly, multiple choice questions can sometimes mislead the student; thus, the issue comes down to when to use them and the form that they should take.

Discussion: To move into the *Discussion* phase, the teacher should simply say something like, 'Okay, let's hear everyone's ideas,' 'Let's share our thoughts,' or 'Let's discuss before the experiment.' The teacher should make every effort to choose a phrase appropriate for the type of problem.

'Let's share our ideas' is probably best used when the problem only requires a vague expectation. In those cases, teachers need not go into *why* those various ideas come up. It is enough to simply note that 'there are lots of different ideas from different people' and then move into the experiment. 'Let's share our thoughts' might be best used when students do have some type of thought or reason as a basis for their expectation, but the points of conflict between students' reasons are not yet clear enough to have a discussion. The teacher may simply note that 'different people think in different ways' before moving into the experiment.

'Let's discuss' should be reserved for cases in which the rationales behind different student predictions are clearly in conflict. Then, the discussion's job is to make clear exactly what the points of conflict are. Once these are clarified through discussion, the experiment will not only demonstrate which prediction was correct, but also what type of thinking (or hypothesis) was accurate.

The statistical reliability of the best phrases to be used for each problem is expected to keep increasing as the HEC Classbook is continually perfected. Nonetheless, depending on the teacher, classroom, or time of the year, students may end up engaging in lively discussion for *Problems* which have not genuinely reached the *Discussion* stage; therefore, teachers do not necessarily have to be bound by the above guidelines. Nonetheless, we hope that they remain conscious that the appropriate level of discussion may depend on the problem.

Lastly, once a brilliant idea brings the points of conflict into sharp contrast, there may or may not be much value in further discussion. At those times, if we still want students to give some consideration to some other idea, we include this in the form of 'Hints' found in the *Discussion* section. The teacher should actively offer the 'Hint' to the class if it has not come up in discussion for some time and the students do not seem likely to bring it up.

Sequence of problems: The concept or theory targeted as the educational goal requires a set of diverse *Problems*. And those *Problems* must be selected and arranged carefully enough so that every student is able to make correct predictions by the last problem in each set of *Expectation* → *Discussion* → *Experiment* cycles. It is a requirement that each and every student be able to make accurate predictions by the end of the problem set.

Although the teacher may struggle over exactly which problems are most appropriate for the principle, if the target concept or law is truly one of the most general, foundational concepts in science, the students will be able to answer any problems correctly. If the teacher cannot find any problems appropriate to assign to students, then the concept or law must be considered inappropriate to teach at that time.

How does a student develop recognition of the targeted concept or law via cycles of *Expectation* → *Discussion* → *Experiment* in a set of *Problems*? We have represented the process in the flow chart shown in Figure 1 on the following page.

The *Problem* section may make use of a segment for new problems entitled 'Let's make our own problems based on what we know.'

In addition to making accurate predictions, incorporating other scientists' ideas, and performing actual experiments, formulating the problem into a format that allows it to be solved is critically important in scientific research. Educators may criticise the HEC for providing pre-made problems in the HEC Classbook and not allowing students the chance to develop their ability to formulate problems. However, in the early stages of science education, students do not have a good grasp of scientific concepts and laws, so it is extremely difficult, if not impossible, for them to create solvable problems on their own. However, after they go through a series of problems and learn a certain concept or law, they may be able to discover new solvable problems. Problem creation can then serve an educational purpose. A section called 'Let's make our own problems' could be added in the appendix of the HEC Classbook, and teachers could choose appropriate problems to present to the class as a whole.

Enrichment Problems and Practice Problems: *Enrichment Problems* are not necessarily intended for the entire class, but only for those students who want additional challenges. *Practice Problems* are problems primarily for review, so they do not have to be proved through experimentation.

Even though teachers should not require all students to try the *Enrichment Problems*, if some students have attempted one, the teacher should have them present their results to the class. This will hopefully encourage students' interest in research and perhaps

Figure 1: Process of developing recognition

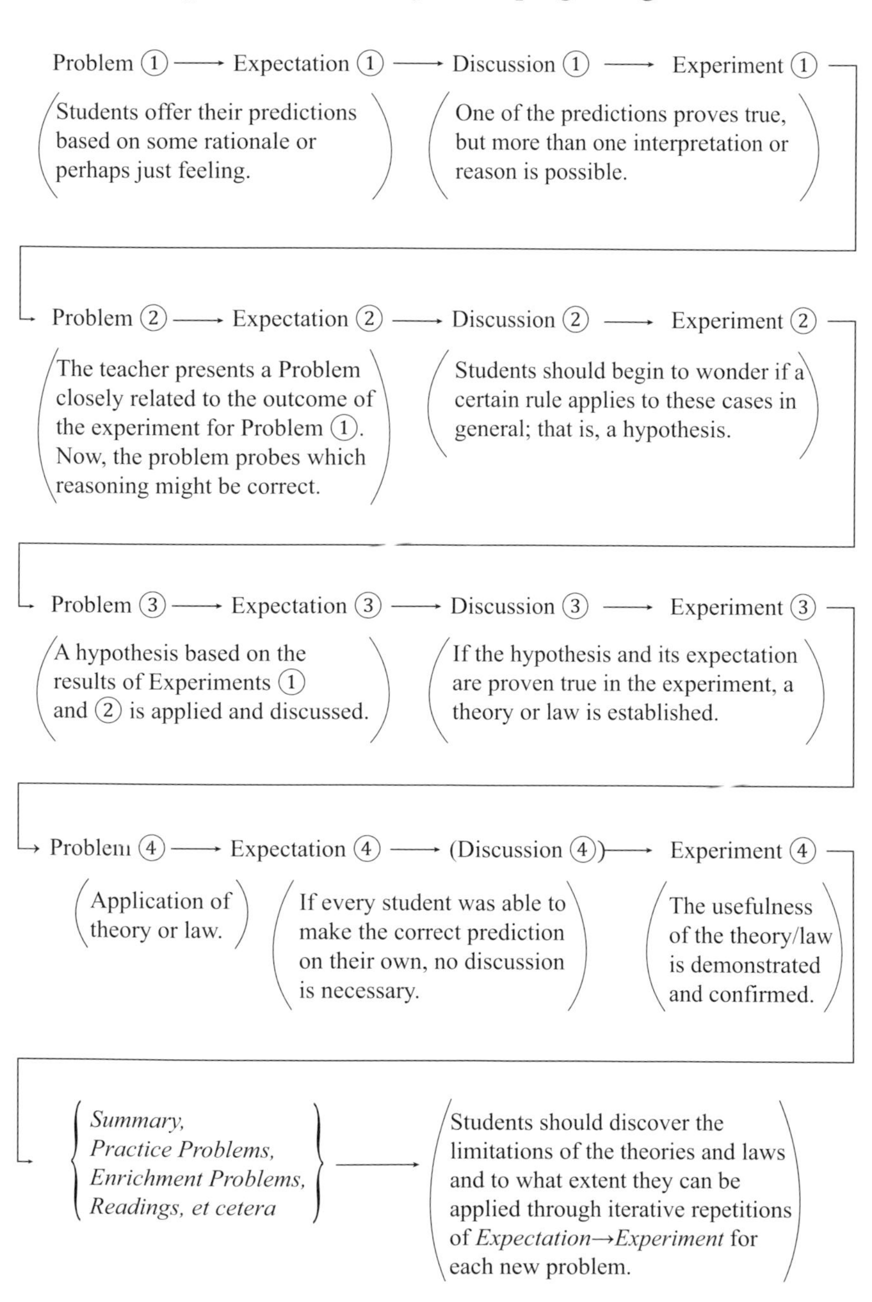

expand their view of the educational materials. Also, if students have a poor showing on *Practice Problems*, then it would be best to demonstrate the answers through experiment. (In general, if any class does not perform well on a problem or seems confused, it should be investigated as a new *Problem* or removed for being beyond the educational scope of the class.)

Definition of terms: *New Science Words* are presented in the course of solving *Problems* in order to make the meaning of certain technical terms clear. The section entitled *Summary (of Theories and Laws)* clearly expresses the target concept only after students have grasped the principle at work.

In the HEC Classbook, the theory or principle at hand is presented at the beginning of a section for some topics and at the end for others. If a theory or law seems to be self-evident or easily intuited (often when the rule is nearly true by definition), the theory or law may be presented at the beginning, so students can actively apply it to coming *Problems*. This is when the rule will be shown to be more effective than an intuitive, commonsense, understanding. In contrast, there are also cases in which the rule or principle is not presented initially, and students are forced to acquire it as a principle by verifying the consistent utility of that hypothesis as they work through many *Problems*. Deciding when to present a theory or principle will require analysis of the nature of that theory or principle and the experimental research on the readiness of the classroom.

Readings: *Readings* present topics which are centred on the fundamental scientific ideas or laws learned in class, with the intention of expanding students' horizons and encouraging their interest and curiosity.

Since the *Readings* will deal mainly with things that students do not directly interact with every day, they may doubt some of what they read; but that is no matter. We simply tell them that things they have learned in class are effectively employed in ways they have not experienced yet and hope they respond with exclamations of 'Really?!' 'Wow!' or 'Whoa, that's crazy!'

2.5 HEC classroom management*

Classroom management for these classes cannot be discussed without talking about the HEC Classbook which supplies the course content. The general principles of classroom management found throughout the HEC Classbook have been developed based on the basic educational theory described in the previous chapter. These principles have been formulated as described below, in association with Mr. Akira Kamisako of Gakushūin Primary School and Mr. Kazuaki Shōji of Seijō-Gakuen Primary School and other incumbent teachers. These principles have already been utilsed in over 100 classes, totalling thousands of class hours, and are recognised as the proper means for accomplishing the three *Goals of HEC Classes*.

A. Introducing the problem

Firstly, students should not prepare for HEC, as it does more harm than good. The problems in the HEC Classbook must not be shown to students before class; rather, each new

---

* Editor's Note: 'The *Kasetsu* Class Album' should be consulted in conjunction with this section, as it charts the steps for HEC classes as set out here in.

section of the Classbook should be distributed during each class session.

The ideal of HEC is achieved when the students think about the problem on their own and choose predictions as they wish. If students know about the problem in advance and prepare for the lesson, the class merely becomes a stage for them to show off and parade their knowledge, learned and memorised beforehand. That is why teachers should follow the distribution rules above.

For this reason, the HEC Classbook as currently structured is compiled one unit at a time by students when they receive the handout sheets for that unit. The students are given a cover file on the first day and then a new handout for each new problem. Students may take their Classbook, which contains the lessons they have studied, home for review.

Secondly, class usually starts with having the students read the HEC Classbook to see what problem they will be examining. The teacher does not have to explain anything but may wish to remind students of lessons from past classes. The HEC Classbook provides all the necessary information to conduct class and has been designed specifically so that the teacher can follow it, without the need for any additional instruction.

Thirdly, in keeping with the Hypothesis–Experiment concept, the *Questions* section should be gone through briefly without getting bogged down in details. For example, the teacher should have the students with first-hand experience or knowledge raise their hands to provide answers.

Fourthly, in the *Problems* section, the teacher should try to get the students to clearly visualise how the stated problem will play out in the physical experiment *before* making their predictions. Once they do this and choose an expectation, the students should then write their expectation in the space provided in the HEC Classbook.

Students should select and write down their first expectation without talking to other classmates. This expectation can be based just on the student's hunch or on careful consideration of previous experience or lessons. Essentially, it is necessary to let the students choose the expectation they think is the most likely. A teacher may even suggest that the student 'just guess' because 'we have not learned about it yet.' It is not important at the beginning for students to have a reason for choosing their expectation, because just choosing something will raise their interest in the experiment's results. Moreover, the HEC Classbook is designed to guide students so that they can eventually make predictions based on substantial reasons. The reason we encourage the students to write their ideas in the HEC Classbook is so that they can develop their own ideas clearly first, and do not simply go along with everyone else. Children who have taken these classes have come to recognise the importance of choosing their own expectations in the beginning: 'I can understand my own idea better.' This process allows the student to make objective comparisons between their own ideas and the ideas of others, and, moreover, to develop those ideas by confronting the results of the experiment.

B. Counting expectation totals and calling on students to speak

Firstly, after all the students have written down their predictions in their Classbooks, the teacher counts the number of students with each choice and writes the totals on the blackboard in a table as in the photo on the following page.

Photo 1: Record of students' first choices

One aim of this process is to ensure all students' participation. At the same time, it shows the students that not everyone's ideas are the same, so they see and appreciate the necessity of sharing ideas and discovering the correct answer through experiment. They will also be able to see where their choice stands among their classmates' ideas.

Generally, totalling students' predictions is a moderately time-consuming task, so using an analyser [automatic voting device] to save time is not unreasonable. However, replacing the activity of a teacher with electronic or mechanical means may not be so desirable, because these prevent the teacher from making eye contact with the students and confirming each of their votes individually; furthermore, the tallying process provides a good point of reference for sharing ideas and for student discussion later. Even if the teacher decides to use an analyser, they should still have students raise their hands, so everyone can see who chose which expectation.

Secondly, after tallying predictions, the teacher has at least one student for each prediction express why they made that choice.

At first, the teacher should call on a student by name (in keeping with educational values) without asking students to raise their hands. Students are allowed to express any reason for their choice including 'feeling'; thus, any student should be able to make some type of statement if called on.

We hope that every student becomes capable of freely expressing his or her candid opinion when asked, 'Why did you choose your prediction?' Therefore, the teacher should make a point of calling on students who do not often express their opinions, thereby providing them the opportunity to improve their communication skills. Additionally, the teacher should instruct the class to listen, and give fair consideration, to any idea regardless of who expresses it (except perhaps when the student has misunderstood the meaning of the problem). Furthermore, the teacher should make it a habit to encourage students who chose unpopular expectations to talk about their ideas. The teacher should also encourage students who chose the same expectations but have different reasons to raise their hand and express their reasoning. However, the teacher should ensure that the time for expressing ideas and reasoning is kept completely separate from the time for debating.

The teacher should allow any student to write down their thoughts in the margins of the Classbook if the student wishes but should not force all students to do this. Also, students who choose to write something down in their Classbook should not be allowed to simply read from it when expressing their ideas to the class.

Writing down one's reason for choosing a certain expectation may help some students clarify their thoughts, but it would just be a time-consuming chore for students who chose an expectation without a clear reason. Hopefully, these thoughts will gradually take shape and become organised during the later discussion stage.

Writing should not be considered essential to the process, so the teacher should not do anything more than suggest that students may wish to write down their idea in the HEC Classbook while waiting for other students to finish choosing a prediction. We should heed students' criticism of conventional science education in which they point out that filling up a notebook is a meaningless chore.

Lastly, if a student just reads from their notes during the discussion, it will be hard for other students to follow and also rob the discussion of its energy, so teachers should be on the alert to prevent this.

C. Managing the class discussion

After each group has presented their rationale for their choice of expectation, the teacher should lead the class into a class discussion so that students who see mistakes in other groups' thinking or feel their own way of thinking is better can discuss those points. The teacher should only call on students who raise their hands and should not force any students to speak (except for special circumstances).

A discussion primarily occurs only between people with clear opinions about the problem at hand, but it still serves a considerable role for others who are just listening. This is because once the discussion clarifies which points are in conflict, an outside observer may be able to see the strengths and weaknesses of both sides. Also, it leads the listener to a more concrete and logical understanding of the differing opinions for the problem. Discussions may also serve a purpose for those who choose an expectation for unclear reasons, by revealing the rationale behind the expectation to them and allowing them to formulate it as an expectation backed by a hypothesis. Therefore, discussions do not necessarily require all or even most of the students to make statements. It is good enough if the discussion is understood by everyone in the class and gets them thinking. Based on the student surveys we have received in classes conducted so far, we can say that students were very happy to have the discussion be a chance for them to think more deeply without worrying about contributing directly.

In contrast, students become very anxious if they are called on and do not have anything in particular to say. This anxiety and experience often mar the student's opportunity to think about the topic while listening to the discussion. Just as scientific research must be independent and agent-oriented, we strongly maintain our stance of respecting the independence and agency of students in HEC Classrooms. In conventional 'education,' students who do not wish to speak are often called on deliberately thus creating anxiety and teaching them to fear external pressure in order to drive them to study. But HEC classes allow students to remain silent unless they have some information that they wish to offer.

Discussions may or may not be very intricate, depending on the nature of the problem and how the lesson has progressed to that point. As already stated, the HEC Classbook indicates approximately what level of *Discussion* the teacher should expect for each problem. The teacher should not expect every discussion to result in a fierce conflict of opinions.

If the discussion does not lead to a conflict of opinions, the teacher may offer one or two questions, hoping to spark some discourse, but they should not panic and try to force a debate. It is best to casually summarise that the expectations have fallen into a few categories and that 'different people have different ideas' before moving into the experiment. These situations often develop into more active discussion eventually, so there is no need to worry.

During discussions, teachers must be quite cautious with their statements, they should not behave in any way that supports only the correct idea or hypothesis. The teacher's statements during discussion should be limited to making sure the whole class understands the content of the discussion, offering supportive statements of minority opinions, energising the discussion and clarifying any points of conflict.

Teachers are generally a presence of overwhelming authority in the classroom and students are extremely sensitive to their teacher's behaviour, so they will try to glean the correct answer from subtle clues. Unless the teacher is extremely conscientious on this point, the students will spend their time and efforts trying to read the teacher's facial expressions and statements rather than listening to the discussion or weighing their own thoughts. In particular, if a teacher attempts to lead the class towards the correct expectation, students will often quickly pick up on this and figure out the right answer before being able to recognise its reasonableness.

Another common issue is when a teacher misunderstands a student's thought process (particularly if the student does not have a clearly organised idea) and rewords the student's thoughts into a different idea altogether when paraphrasing for the rest of the class. Teachers who are new to HEC and do not have a strong understanding of how students might generally think are particularly prone to misapprehending students' statements. Those teachers often do not realise that there is more than one way to think about a given problem and are apt to see children's ideas as too unreasonable. Teachers should be careful not to misunderstand students repeatedly because this will cause discussions to become overly confusing. In order to avoid such misunderstandings, the teacher should avoid bold rewordings of a student's statement and should even consider having other students attempt to paraphrase the idea. If the initial student's statement has some measure of logic to it, other members of the class will probably be able to offer a clearer interpretation of it.

To make discussion livelier, teachers should present unpopular, unique and unconsidered opinions, and steer the discussion towards an examination of any points of conflict. These actions are very effective for improving the impact of discussions. However, in order to do this efficiently, the teacher must have deep knowledge of the problem and discussion at hand, so we do not expect most current teachers to be able to do this right away. If a teacher does not have sufficient understanding of the core concepts associated with the problem and attempts to delve in too deeply, the main idea might get thrown by the wayside, with a net negative effect on the class. We, therefore, consider it wise for teachers to avoid speaking during discussion. We are presently working on a teacher's edition of the HEC Classbook which will contain exhaustive explanations to help teachers make the most effective statements.

D. Accepting students' changes of expectation

If a student makes a request to change their choice of expectation during or after the discussion, the teacher should accept the change and make the necessary corrections to the totals, as in the photo on the next page.

Discussions are given emphasis in HEC classes because they help students rethink their ideas about a problem and summarise those ideas into a more organised hypothesis. This occurs through the process of listening to other students' ideas, accepting those

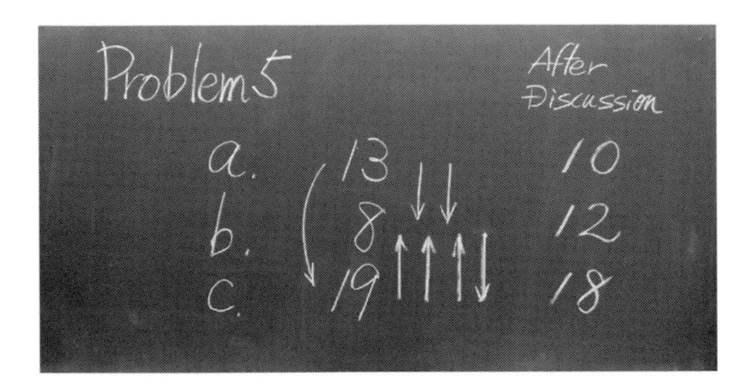

Photo 2: Shifting student choices

ideas or finding arguments against them, and stating one's own ideas, convincing others of their validity or dealing with criticism. Therefore, it is only natural that, during a discussion, students may decide that their initial expectation was wrong and that they would like to change it.

However, in the early stages of our classes, we found many students who thought it was 'cowardly' to change their initial expectation despite realising during the discussion that it was wrong. These students would stubbornly sit with their expectation, though they knew it was wrong. At times like this, the teacher should make every effort to help the student see that understanding and adopting the good aspects of others' ideas is essential in science and that changing one's opinion is no cause for shame.

Whenever a student wants to change their opinion, the teacher should ask the student to explain their reasoning to the entire class: 'Could you tell us why you would like to change your expectation after hearing so-and-so's thoughts?' This may help the discussion to hone in on the points of conflict.

In HEC class discussions, it is not always true that the group with the correct expectation also makes the best arguments or has the most support. It may happen that one or several groups with a wrong prediction make the most compelling arguments in the discussion, winning over many students. But this should not be any cause for alarm for the teacher.

Many educators may worry that if the expectation and rationale which was the most logically organised and convincing in the discussion turns out to be falsified by the experiment, and that a less logically organised expectation turns out to be correct, students may learn to distrust logic or become disheartened with learning in general, creating a severe roadblock for their general education. However, we should not cave in to this brand of 'almighty logic' rationalism.

Science does value logic and rationality, but logic is not all there is to science. The history of science provides us with many cases in which a logically robust theory, one completely consistent with all experiments up to that point, made predictions that were then flatly falsified in a new experiment. Aristotle's theory of motion has long been one such example, followed by the more recent Michelson–Morley experiment and blackbody radiation experiments. However irrational the results seem to be, the factual reality of the experiment tells us that a new theory must be developed. Therefore, rather than avoiding this situation, educators should actually embrace the opportunity for students to experience the reality that sometimes the most plausible theory is proven wrong by experiment. This will demonstrate the importance of experiment and the limitations of reason in science.

This is, however, an issue of degree. If students repeatedly see their most plausible ideas overturned by experiment, they are bound to lose confidence in their ability to think rationally. But with the HEC Classbook, once the students have discovered the basic target concept and principle in the first few *Problem* → *Experiment* cycles, they will be able to get the right answer for the remaining problems by using the concept and thinking rationally. This helps us avoid the 'almighty experience' empiricism of segmented learning in everyday life, which ignores the principles, rules and concepts fundamental to science.

Since the teacher does not have to worry about keeping the group with the correct expectation 'in the lead' during the discussion, they should focus on guiding the class to clearly see what the points of conflict are. (In this way, when students see which expectation was correct through experiment, they may also recognise what type of thinking or ideas were not true.)

Once the discussion has reached a lull, the teacher should propose that the class move into the experiment, by saying something like, 'Well then, shall we see which expectation or idea is correct by doing an experiment?' At this stage the teacher should ask the students for 'Any last thoughts?' or 'Anybody who wants to change their choice of expectation before we do the experiment?' to wrap up the discussion.

It may be quite a challenge to judge when to finish up with the discussion and head into the experiment, especially during lively discussions, but the teacher should make it a rule to respect the general consensus of the class. At times, a smaller group may be engaged in a very abstract or abstruse discussion that the majority of the class cannot follow. The teacher may have to interrupt the discussion and ask the class for a vote on whether to move into the experiment or keep discussing. However, individual teachers will have a range of various pedagogical viewpoints from which to judge how much time to spend discussing a given problem.

In HEC classes, highest praise should be reserved not just for students who select the correct prediction but also for those who play an active role in the discussion. The teacher should encourage the students to recognise this on their own as well.

For the first few problems, students are simply being assigned something that they have never been taught before, so they will rarely come up with the correct guess and reasoning. Getting the right answer at the beginning is not at all important. Rather, students should be praised far more for being able to call upon all of their experience, imagination, and reasoning ability to come up with an expectation and muster convincing arguments in the discussion, even if their expectation ends up being wrong. This type of activity helps lift what was merely a difference of expectations into a difference of hypotheses, which encourages a deeper reading of the experiment to come and will assist others' development of recognition. Furthermore, such abilities are extremely important in scientific research.

Also, although it is important for a student to be able to apply the central idea or rational thinking learned in cycles of *Expectation → Discussion → Experiment* to correctly answer new problems, this is not everything. As we have already stated, simply knowing the right answer yourself does not constitute scientific knowledge. You also must understand the issue well enough to be able to convince others that you are correct. Therefore, if you have grasped some central concept or principle before your peers, you must now make the effort to convince them that your thinking is correct. There are some students who instinctively grasp this importance without being given special instruction from the teacher. However, there are also many students who only value their individual ability to make correct predictions and do not give much thought to improving the discussion or the group's collective recognition, so the teacher should be on the lookout for this kind of situation. HEC classes attempt to maintain the social nature of scientific recognition even in these situations.

E. Performing the experiment

Experiments must always be used to verify class expectations, except perhaps in cases when every member of the class has the same, and furthermore correct, expectation and every member agrees that there is no need to perform the experiment.

Once a class has cycled through a few problems and all members have grasped the central concept or principle at work, the whole class might come up with the same prediction for a new problem with all members in agreement that there is no need to do the experiment. In those rare cases, the experiment may be bypassed. However, if even one student would like to verify their expectation through experiment, or is not quite sure, then of course it is necessary to do the experiment.

It might be a good idea to include a measure of how much confidence students have in their expectation before the experiment. It could be a survey which asks: 'How confident are you in your prediction?' The answers could be: '100 per cent sure,' 'pretty sure,' 'not sure' and 'not sure at all.' When a student's expectation is correct, there will be a big difference in its value depending on whether they were '100 per cent sure' or not. Furthermore, if a student's expectation was wrong, they will probably have to change their thinking a lot more if they were '100 per cent sure' than if they were wrong and 'not sure at all.'

If everyone in class has the same expectation, but the expectation is wrong, the teacher should simply express some doubt about the students' expectation, support a different one, and then perform the experiment.

Experiments must be performed with only enough precision to clearly determine which expectation is correct.

For most experiments, *precision* must be taken into consideration. For example, when you attempt to measure the force of a spring by its stretch, the spring will not follow Hooke's law perfectly. So, if you do not give enough consideration to how precisely that spring will be able to reflect the strength of a force and just rush in to measure its stretch, the results might not only be worthless, they may lead to significant confusion. For instruments only good down to the centimetre level, you cannot measure down to the millimetre level and then complain about inaccuracy. To offer another example, even though books tell us that the boiling point of water is 100°C, we may get a two to three degree range of error due to the sloppiness in the experimental setup or technique. It would only cause confusion to worry about whether to read the thermometer as 98°C or 100°C.

However, it is not easy, in general, to determine theoretically or experimentally how much error we can expect for any type of experiment. Therefore, before that issue comes up in the experiment, we should make sure that our final measurement is not more precise than is necessary to check students' predictions. For example, when we are predicting whether a spring will stretch 5, 10, or 20 cm, the experimental results can be read roughly; that is, we do not have to read the measurement of the spring's stretch as 5.5 cm instead of 5.0 cm. In fact, this more accurate reading should be avoided, as there is no problem with ascribing the difference between the expectation and the measured value to experimental error. Reading the gradations more finely than the error range allows is not more precise but actually imprecise. (Many educators claim that this would be an ideal time to help students discover the secondary causes leading to experimental error. However, if that is not delayed until after the primary educational goal [i.e., concept, hypothesis and confirmation], it will just lead to confusion. This is, thus, a topic best avoided in general.) Therefore, we can say that reading gradation lines and other measurements in these experiments should be done roughly, just enough to judge which of the competing expectations is right. Also, the students themselves will have just finished making their guesses and debating, so they are usually just concerned with which guess and thinking were right. They are not looking for anything more, so rough measurements are sufficient.

There are some experiments which all students must be allowed to perform individually for themselves or else they may doubt the objectivity of the experiment. However, in other cases, the teacher or a student representative may perform the experiment in front of the class. However, if anyone has any doubts about the experiment, it must be repeated as many times as necessary.

The crux of an experiment is holding some specific expectation regarding an object or phenomenon and checking that expectation against reality; whether the experimenter is physically part of the experiment or not is immaterial. If the students each make a firm prediction before the experiment, and do not find any cause for doubt once the experiment is performed by the teacher or another student, there is no need for each one of them to perform the experiment. Educators often bring up whether or not a student who performs an experiment using their own hands is far more likely to remember it than a teacher-led experiment. But if the student has their own prediction in mind and carefully observes the experiment, even a teacher-led experiment will leave a strong impression. Furthermore, if the experiments were to be performed by every student, not only would there be added cost and time, but without strict controls, students may play with the equipment before coming up with any expectations, or possibly even discover the principle before giving it any thought. That is why for the early stages of HEC classes, we advise against student experiments. Some students may not yet have a good idea of what an experiment really is; they may simply toy around with the experiment equipment without any ability to conduct any inquiries into natural phenomena.

On the other hand, there are times when the students may not be able to accurately see the experimental process or results, or they may doubt the teacher's experiment, until they are allowed to perform the experiment for themselves. Naturally, student-run experiments are appropriate in these cases. Also, though not necessarily emphasised in HEC classes, experiments which aim to foster a minimum level of experimental skill in the student should naturally be performed as student-run experiments. Moreover, if a student wants to confirm a teacher-led experiment with their own hands, giving that student the opportunity is a good idea.

Once the experiment is over, the teacher should simply confirm with the students which expectation was right, which ones were wrong, and what ideas in the discussion were right or wrong before moving on to the next problem. The teacher must not push any new interpretations for the experiment on the students.

If any students offer up new interpretations which were not offered during the discussion for the experiment result, the teacher must be careful to simply clarify the idea and state that it would have to be examined in a different experiment. In the HEC Classbook, there are often *Problems* later in the unit which investigate those new interpretations. Thus, the teacher can simply leave the new interpretation for checking in the later *Problems* section. However, if there are not any *Problems* for that interpretation in the Classbook, and the interpretation is worthy of examination, the teacher should attempt to create an experimental problem on the spot in order to test the interpretation. We also ask that teachers inform the authors of the HEC Classbook of any new interpretations and results. These situations can certainly lead to further improvement of the Classbooks.

F. Expanding the topic and implementing tests

The HEC Classbook is designed so that each student will be able to make the correct prediction by the final problem in any given unit. When this goal is not achieved, the teacher should (repeatedly) attempt to create similar questions extemporaneously to help all students make the right prediction. We also ask that teachers report the details of such cases to the authors of the HEC Classbook.

Although we say 'every student,' we of course may make special exceptions for students who missed previous classes as well as intellectually challenged students who are generally unable to participate fully in classes. Also, if a few students came up with the wrong expectation (excluding careless mistakes), the teacher should add on an additional problem or two. We are currently planning to develop such extra questions for HEC classes; but, theoretically, if the class does not achieve the goals in the time set out in the HEC Classbook, this means either that there has been a mistake in the running of the class, or that the HEC Classbook must be revised. It would thus be best to contact the authors in such cases.

The explanations in the *Summary* and *New Science Words* sections are only to guide students to understand the *Questions* and *Problems* and do not need to be completely understood.

Some *Summary* and *New Science Words* sections might seem relatively easy to understand at first glance but are actually very difficult to apply to specific problems. However, as long as the student comes to understand those items once they become relevant to a problem in the HEC Classbook, there is no need to worry. It may be too confusing to try to explain how everything in the *Summary* and *New Science Words* sections might apply to some later problem, so any such explanations are best avoided.

The procedure for handling *Enrichment Problems*, *Practice Problems* and *Readings* was set out above in the 'How the HEC came to be' section, so we will not repeat that information here.

Tests are not necessary in HEC Classes. However, tests may serve a purpose if the teacher wants to give students the opportunity to confirm their progress or boost their confidence. If the teacher does decide to give a test, they should be careful to create proper test questions that demonstrate new applications of the material. If several students had trouble with a certain question, it should be examined through an experiment performed in class.

There are tests that evaluate the efficacy of the teacher or the HEC Classbook, those that evaluate student ability, and also tests that can serve both of these purposes. In HEC classes, the tests for evaluating the HEC Classbook or classroom are designed to make sure that goals for the class are reached (as has already been stated), but these tests can also evaluate student ability. Most students are not likely to be satisfied with their ability unless they get over 80 per cent on tests for recently covered material; therefore, we have set the standard at 90 per cent. A class average of 90 per cent means that the majority of students got between 80–100 per cent, with a small number of students in the 60–80 per cent range. Although students who get 60–80 per cent will not necessarily be the most confident, they may be at least a little pleased with themselves if they are normally under-achievers. In any case, tests should never become something that hurts student confidence.

The teacher must be conscientious when creating test questions. If more than a quarter of the class cannot solve a test problem, the problem probably contains some new element beyond the scope of the original concept. If the problem goes to a level not taught yet, students do not need to be able to solve it outright; it is a problem that they should hypothesise about and verify through experiment, so the teacher should remove problems like this from the test and treat them as a hypothesis problem to be checked via experiment.

Also, there may be problems which largely rely on keen insight or a so-called 'eureka moment.' Such tricky eureka problems are heavily reliant on the serendipity of the moment, so 100 per cent completion should not be the goal. The teacher should view a score of two out of five on these problems as meeting the standard. Science as practised in society makes use of insights from a large number of people, so it is not necessary for one person to think of everything.

3 Hypothesis–Experiment Class as Democratic Education

The HEC was originally designed to help students gain a solid grasp of some fundamental scientific concepts and was not proposed as a class with any direct connection to democratic education. However, once it was actually put into practice, we discovered that these classes had effects far beyond what we had first intended. We realised that not only did the students gain a definite understanding of basic science concepts, but they also learned scientific thinking and attitudes, how to think creatively, and even developed a strong sense of what democracy is.

Of course, this is not necessarily a complete coincidence. We believe that science cannot be taught without getting the students to think with scientific creativity. Also, we think of science as both the process and result of a mutual sharing of ideas, mutual criticism and creating a theory that any person would be able to accept. And since we have built the framework of HEC classes on these beliefs, it is natural that our classes should have a deep connection to creative thinking and democracy. However, it was hardly our expectation that the classroom method itself would lead children to 'learn how to learn.'

By listening to what students say in HEC classes, observing their thinking patterns and actions, and reading their comments in the after-class survey, we can see that they have clearly gained an understanding of creative thinking, scientific recognition and even democracy in the course of these classes. On this basis, I would say that HEC classes are now producing unexpected results. I would, therefore, now like to briefly describe my thoughts regarding what meaning HEC has as a form democratic education. I do this because I am not sure people will understand the significance of 'democratic education' from just seeing the words.

One thing that will be deeply impressed upon students who take HEC classes is that 'truth is not determined by the majority.' Through months of class, there will be at least a few times that a small group makes the correct prediction and the majority of the class makes the wrong prediction. There will also be times when only two or three people in thc class come up with the correct prediction. Thus, children will keenly realise that simply because their opinion is shared by the majority, they cannot just relax, assuming that they are right, or make fun of others' opinions.

The principle of the majority may be one of the fundamentals of democracy, but this principle includes an understanding that the minority may at times be correct, so unpopular or minority opinions must also be respected. Respecting unpopular opinions is the most critical aspect of democracy, for if one does not have a proper understanding of this, then democracy falls into either mob rule or its polar opposite, fascism. However, HEC classes teach this delicate, most difficult aspect of democracy in a captivating way. The overwhelming strength of these classes is that they show that truth is not determined by the majority or by the teacher, but by the natural event itself – that is, the experiment. It is certainly the case that the majority will often be right, because once most students develop sufficient logical insight into the problem, most or all of them will find the correct prediction. The children who take this class will develop the insight to help them discern when the majority opinion is likely to be correct and when a minority opinion might be correct.

Discussion, so fundamental to democracy, is only established when there is respect for unpopular opinions and the consideration that such opinions may be right. This provides some clarity for why discussions are so well received in HEC classrooms.

The good thing about democracy is that anyone can come up with a brilliant idea and everyone's ideas can be used; thus, anyone can contribute to the group's development. Schools often have some type of ranking from remedial to top students, but that is merely an 'on average' determination. Even the lowest-performing students may come up with superb ideas at certain times and places.

That is the nature of society. One of the principles of equality in democracy is that all people are able to contribute to society in one form or another.

This concept also seems to be deeply impressed upon children in HEC classes: advanced students do not always make the right prediction. Sometimes, a student whose class performance is normally underwhelming ends up taking on the whole class, stands firm with a powerful argument, and is then proven to be correct in the experiment. At times like this, children fortuitously gain the understanding that the group moves forward by appreciating and respecting human individuality.

For the most part, in HEC classes, slow students do not get left behind and advanced students do not get bored. This is because the classroom and the HEC classes, respectively, mirror society and the development processes of democracy. Advanced students are expected to teach the slower students, but it is not moral reasoning that underpins this aspect of classes. Offering an easy-to-understand explanation that slower learners can comprehend also enhances the development of children who have already figured out a problem. The children will gain a sense for this without being taught it directly. Explaining something to someone in an easy-to-understand fashion first requires one to understand one's own thoughts sufficiently, and children will learn how extremely valuable that is. In society, smart people are not an elite; rather, people who can give others a hand up (in a variety of senses) are considered great. Thinking about how great individuals emerge in society helps us to see the social importance of this type of group learning.

There are still many things I'd like to say about the relationship between HEC and democratic education. Unfortunately, I do not have enough time to expound on this in detail; therefore, I would be satisfied if you were to read these comments as a brief memorandum on my ideas. I would simply like to add one more thing. That is, democratic education is not so much presenting a definition of democracy, describing the attitudes fit for it, and then discussing it. Instead, democratic education is alive when the class itself is carried out in a democratic fashion. In this sort of class, students will have an even keener sense than the teacher of what democracy is about, even though they may not be able to express it in their own words. Humans have always been social beings, and generally want a sort of democratic system of cooperation, so they do not need a sermon on democracy.

4 Memorandum Regarding Hypothesis–Experiment Class

Learn from children's mistakes

Always remember that when children make a mistake in judgment, it is not because they are ignorant; it is because they are perceptive. It was the presence of particular 'characteristics' in the problem that led them to make a mistake. Because they are perceptive, they noticed these 'characteristics' and consequently made the mistake.

It is thus the height of folly to mock or chide students for making a mistake. The teacher may not know why the student made a mistake, but the student does, so the student has more knowledge than the teacher in this regard. At times like this, the teacher must learn from the student. We must learn what type of experience the child has and what associations or ideas they came up with based on their experience. Often, in these cases, the student will have thought of the teacher's specific question as secondary to the issue.

It is truly foolish for teachers to mock or chide students. Equally, one who presumes to be a teacher of teachers also must not mock or chide those teachers. A teacher of teachers must also learn.

What to teach

In education, we should not give too much emphasis to things that can be understood at a glance or by doing one simple experiment.

That sort of knowledge can be obtained whenever necessary by doing the experiment or seeing the results once, so it can be left for those times. We generally call this sort of knowledge 'specialised knowledge,' and it is the sort of knowledge that can be gained somewhere other than through the organised, systematic education available at school. However, anything that cannot be learned quickly enough when it is needed (even some specialised knowledge) must be rigorously taught in advance through education.

The basic concepts and principles found in HEC classes are not this type of easily acquired knowledge. They are concepts and principles that cannot be grasped in one experiment or at a glance. This is largely the case with the fundamental concepts in science. Even though we tell students that the concepts and principles which apply generally to a huge range of phenomena are indeed 'true within a wide range and not valid beyond that range,' just showing them one instance or doing a lone experiment clearly cannot be enough for them to truly understand.

And yet, elementary school science education up to this point has basically consisted of just teaching children things that can be understood with one experiment. Thus, their education ends up just being a potpourri of fragmented, miscellaneous points, which are not very compelling. Worse, in conventional education, even some things which could never be understood with a single experiment have been taken up in class and nonetheless taught as something to be learned via one experiment or at one glance. This has led to abysmal results. Piece-meal knowledge can be obtained piece-by-piece via single experiments, but children have no chance of understanding deeper concepts this way. Then, without ever considering that there might be a problem with the way we were teaching, many educators jumped to the conclusion that children are simply unable to understand these basic, general concepts of science. They then came to the decision that it would be better not to teach any principles (or theories) in science education.

HEC classes emerged from criticism of these decisions. We strongly maintain that principles which cannot be understood through a sole experiment or observation must be examined in a variety of settings and tested in a variety of conditions. That is why the ordering of problems – experiment or observation – is extremely important, and also why the creation of the HEC Classbook was so important. One-shot one-experiment education stops at the level of 'an expectations and experiment class' but to make a deeper level of enquiry possible you have to have stages of 'expectation–hypothesis–discussion–experiment.'

Is learning enjoyable, or not?

Knowledge is power. Science gives us the desire to imagine and face the unknown. Gaining new knowledge is enjoyable but learning things that are never useful is meaningless and not fun; it is far more enjoyable to acquire knowledge for which we can see wider applications.

Even isolated or individualised knowledge, if sufficiently useful, is fun. For example, we learn the alphabet, common words, the names of friends we see often, the names of plants, animals, and things that people talk about. But if we make a false step here and force children to memorise the names of plants and animals that are not commonly used, enjoyment takes a nose dive. If the knowledge in your possession does not have a useful outlet, it merely exists for you to show it off.

General knowledge with predictive power over many things is far more than enjoyable to learn than isolated knowledge about individual things. Not only is such knowledge very useful on its own, it also allows you to expand your own knowledge through the work of a richly creative mind (heart). Knowledge that lets you guess 'I bet it's like this' and be right is a knowledge full of dreams.

Moreover, scientific concepts are powerful enough to make consistently true predictions which may even betray our intuitive understanding. They surpass the joys of intuitive imagination and take us into the realm of the joys of theoretical imagination. We could say our perceptive horizons have opened up, allowing us to make far bolder predictions and explore the unknown.

Do not demand perfection from humans – aim for ninety per cent

Humans should not seek perfection. For example, just because we have taught a child multiplication, we should not expect them to get all problems perfectly right. Humans make mistakes; we should allow for people to have minor lapses. I have come to firmly maintain that it is when perfection is not expected of humans that their nature as free beings, not slaves or machines, is most plainly visible.

If a student misses more than ten problems on a 100-problem multiplication quiz, that is good evidence that either the student does not understand multiplication too well or was not in good physical or mental health. But what can we say if a student misses one or two problems? Is that not simply a human making a forgivable human error?

And is it not precisely because humans cannot avoid mistakes that they are able to discover the unknown and things they had not previously imagined?

Education that demands perfection from humans – education that demands 100 per cent on tests – is an oppressive, militaristic education. It is inhumane education that will turn students into slaves. We should not be demanding perfection from humans or students. I have heard that when typists or key punchers are given a test in which they are carefully supervised for mistakes, they get too nervous and end up making more mistakes than usual. This suggests that people are people and they do their most efficient work when working at their own pace. We should not put children under constant stress trying to guarantee there are 'no slip ups.'

That is why I decided to be a *ninety per cent-ist*. If one or two children completely ignore class, it is okay; they probably have their own situation or reasons. We should be worried if the same child is always ignoring class, but if it is just that 'my cousins are coming over to play today and I haven't seen them for a long time,' it is all right for that child to be daydreaming during class. A class that is too strict on those children is part of a slave-to-anxiety type of education.

We must maintain the goal of getting every child to understand, getting them interested in science and in learning about the unknown and in how to think generally. However, we must not demand that fully 100 per cent of children fulfil our expectations according to the end-of-class survey. Yes, it may be a problem if the same child says they do not like any class, but if a child just happened to not enjoy some aspect of class, it is better not to worry. The teacher will lose their own freedom if they start to worry about such things, thereby forgetting something vastly more important.

We should be satisfied reaching the ninety per cent level for our goals. As long as it is not always the same children in the remaining ten per cent.

First Appearances and Research History

The first appearances of the four articles, in Japanese, included here are as follows.

1. 'The Process of Establishing Mental Recognition in Science,' *Rikakyōshitsu* (The Journal of Science Education, Japan), *1966, June.*
2. 'What is the Hypothesis–Experiment Class?' *Rikakyōshitsu* (The Journal of Science Education, Japan), *1966, September–October.*
3. 'Hypothesis–Experiment Class as Democratic Education,' *Seijō Gakuen Elementary School Bulletin, No. 2, 1964, October.*
4. 'Memorandum Regarding Hypothesis–Experiment Class,' *Hypothesis–Experiment Class Research, No. 8, 1966, December.*

The source of the selections and quotations used in this document is Dr. Kiyonobu Itakura's *Science and Method* (Kisetu-sha, 1969), compiled by Shigeru Nakahara. We will touch on the first appearances of each of the four selections, as well as Dr. Itakura's research history, up to and including his proposal of the Hypothesis–Experiment Class (HEC), drawing from the 'Afterword for Young Readers' in his *Science and Method.*

Dr. Itakura entered the University of Tokyo, College of Arts and Sciences (Natural Sciences Stream 1) in July 1949. His was the inaugural university class under the new Japanese school education system implemented after World War Two. He is said to have applied after seeing an article in the newspaper stating that coursework in science history would be offered at the new college.

In his third year, he proceeded into the Department of History and Philosophy of Science. However, not one professor there was researching the history or philosophy of science. Dr. Itakura was sorely disappointed with this, so he barely attended classes, instead choosing to form a group to research natural philosophy and the history of recognition. He also formulated his own research topics within science history, based on his readings and discussions with other students. His research on the history of science began with the heliocentric and geocentric theories of the solar system, moving into the history of classical mechanics, electricity and magnetism, and quantum mechanics.

In the 'Afterword', Dr. Itakura states that he 'learned a lot from student and peace movements in university. I first learned the true meaning of what an experiment was through participating in such student and peace movements at the time. There had never been a time from elementary school to university when I had learned what an experiment was really about, at least not in any deep way.'

In March 1953, Dr. Itakura entered the University of Tokyo, Graduate School of Mathematical and Physical Sciences to research the history of physics. He chose the history of classical mechanics and electromagnetism as his area of research, publishing papers like 'A Comparative Study of the Establishment of Classical Mechanics and Electricity & Magnetism' in the journal of 'The History of Science Society of Japan', *Kagakushi Kenkyū*. In 1958, he submitted this series of papers for his dissertation and received his PhD.

After taking his degree, Dr. Itakura obtained employment at the Laboratory of Science Education in the National Institute for Educational Research. Dr. Itakura had already come to view science education and science history as the same thing from the perspective of establishing scientific recognition. And he was deeply interested in the problems of education, because they offered the possibility, as it were, of doing experiments, which is impossible in the field of science history. Thus, even though Dr. Itakura majored in science history, this job was not exactly foreign to him in any way.

At the National Institute for Educational Research, Dr. Itakura began researching the technologies and social conditions that existed in Japan from the Warring States period to the Edo period, as a comparative study to Europe and its history of scientific research. He was interested to discover 'why science didn't develop in Japan but did in Europe.' He came to the conclusion that, 'Japan did not have the logical philosophical model that one could label an "atomism." Europe, on the other hand, had had that idea since Ancient Greece.'

For Dr. Itakura, this conclusion carried with it a critical hint for revolutionising science education. With the assistance of active teachers, he began the process of reforming science education, which is to say, he began HEC.

Dr. Itakura first proposed a specific class plan called 'A Text for Hypothesis–Experiment Class: *Pendulum and Oscillation*,' which should be considered a prototype of the *Jugyōsho* (Hypothesis–Experiment Classbook), at the National Conference of the Association of Science Education in August 1963. This should be considered the time when Dr. Itakura first proposed HEC.

On November 8, 1964, shortly after HEC was proposed, the event titled 'Science Education in Elementary Schools Centred around Hypothesis–Experiment Class,' was held to publicly show actual classes at Seijō Gakuen School in Tokyo. '**Hypothesis–Experiment Class as Democratic Education**' was written as the preface for the Seijō Gakuen Elementary School *School Bulletin* at that time. This article was written in easy-to-understand spoken language unlike the dense sentences of the two original papers mentioned below.

After initially proposing the idea of HEC, Dr. Itakura moved away from academic research in science history and devoted all his efforts to researching this new HEC. For Dr. Itakura science education research was the touchstone for his theory of recognition and his research in science history. He continued publishing his findings, with *Rikakyōshitsu* publishing both '**The Process of Establishing Mental Recognition in Science**' (June 1966) and '**What is the Hypothesis–Experiment Class?**' (September and October 1966). These two articles are his original thesis advocating HEC, as they include detailed descriptions of the theory and method, and most of the relevant issues.

HEC classes, in comparison to conventional school education, feature innovative theory and methods which can be difficult to grasp. In these two articles Dr. Itakura took pains to fully and meticulously describe the idea, often sacrificing readability in the final document. In the 'Afterword' he explains 'these articles feature a vastly different style than my previous works. This is because I purposely started to imitate Newton's *Principia* in the course of writing them.'

Rikakyōshitsu is a monthly journal first published in 1958 by the Association of Science Education (founded 1954) and intended for a general audience. The Association of Science Education is a voluntary group of individuals interested in science education studies, made up of teachers from every level of education, researchers, students and laypersons. Dr. Itakura was an early member.

In addition to proposing the idea of HEC, Dr. Itakura formed the Association for Studies in Hypothesis–Experiment Class (ASHEC). He also issued *Hypothesis–Experiment Class Research*, a research journal published sporadically as a means of encouraging research into HEC. It included classroom plans, reports from classes actually conducted, as well as some of Dr. Itakura's articles, and even his '**Memorandum Regarding Hypothesis–Experiment Class**.'

ASHEC

Dr. Itakura, along with early supporters and practitioners, organised ASHEC in order to make HEC a reality. As of 2016, there are over 1,200 members from all across the country. The members create, revise and propagate Classbooks and related teaching materials, as well as hosting training sessions, talks and seminars for beginners, both in local areas and nationally. All activities are voluntarily proposed by individual members, who are then organised into an executive committee or other operations groups.

Dr. Itakura had been the spokesperson for ASHEC until just before his death. Currently Saburo Takeuchi is acting as the ASHEC spokesperson. The office of ASHEC is run by Kiyokazu Inuzuka in Aichi Prefecture.

'*Kasetsu Jikken Jugyō Kenkyūkai* NEWS' is an internal publication, consisting of articles written by members, that is published monthly. '*Tanoshii Jugyō*' is a monthly magazine (first published by Kasetu-sha in 1983), overseen by an editorial committee made up of ASHEC members, which has garnered a wide following of not only educators but also many laypersons as a public education magazine.

The following URL is for ASHEC's portal site, created on a volunteer basis by the members. There is also an English page available.
http:/www.kasetsu.org/

Further Reading

Dr. Itakura's Principal Works

In Japanese

Science and Method. Kisetu-sha Co., Ltd., Tokyo, 1969.

Science and Hypothesis: The Road to Hypothesis–Experiment Class. Kisetu-sha Co., Ltd., Tokyo 1971.

Imitation and Creation. Kasetu-sha Co., Ltd., Tokyo, 1987.

History of Science Education in Japan (Appendix/Chronology). Dai-Ichi Hoki Co., Ltd., Tokyo, 1968, Appended 2009.

Hypothesis–Experiment Class: Put into Practice with 'Springs and Force.' Kasetu-sha Co., Ltd., Tokyo, 1979.

In English

Kiyonobu Itakura, and Shunichi Kubo, 'The understanding of the fundamental ideas of physics,' *Research Bulletin of the National Institute for Educational Research* (9): 1–20, 1967.

Kiyonobu Itakura, 'The Hypothesis–Experiment-Instruction method of learning,' *Research Bulletin of the National Institute for Educational Research* (13): 5–16, 1975.

Primers for Teachers

In Japanese

ABCs of Hypothesis–Experiment Class: An Invitation to Fun Classes. First Edition. Kasetu-sha Co., Ltd., Tokyo, 1997.

The Future of Science Education. First Edition. Kokudo-sha Co., Ltd., Tokyo, 1966. Reprinted by Kasetu-sha Co., Ltd., Tokyo, 2010.

'*Tanoshii Jugyō*' Editorial Committee, *Let's Start Doing Hypothesis–Experiment Class!* First Edition. Kasetu-sha Co., Ltd., Tokyo, 2008.

Main Classbooks

Initially, most of the HEC Classbooks were written for the physical sciences. However, in the fifty years since their inception, a wide variety of other fields have been covered by Classbooks as well. Currently, there are over 200 HEC Classbooks and drafts (HEC Classbooks which are still undergoing revision) being created, and they are continually put into practice and reviewed.

Explanatory manuals – created by members of the Association for Studies in Hypothesis–Experiment Class, with significant practical experience in the classroom management and experiment procedures of HEC Classbooks – are also available for many of the following HEC Classbooks.

Natural Sciences

Air and Water
Animals with a Spine
Batteries and Circuits
Buoyancy and Density
Combustion
Crystals
Dissolving
Electric Current
Temperature and Molecular Motion
Flowers and Fruits (seeds)
Food and Ions
Force and Motion
How Many Legs?
If You Could See an Atom
Let's Play with a Handy Generator
Let's Play with Dry Ice
Light and Magnifying Glasses
Limestone – A Mysterious Rock
Living Things and Cells
Living Things and Species
Magnets
Microwave Ovens and Electromagnetic Waves
Moon, Sun and Earth
Objects and their Weight
Pendulums and Oscillation
Pulleys and Work
Rainbows and Light
Seeds and Sprouting
Springs and Force
The Road to Space
The Surface of Water
The Three Phases of Matter
Torque and Centre-of-Gravity
When You Find Free Electrons

Mathematics

Figures and Angles
One and Zero
True Numbers and Fake Numbers
The World of Falling Motion
Space and Area
The World of Doubling and Tripling

Social Sciences

Flags of the World
Introduction to Japanese History
Countries of the World
Money and Society
Prohibition and Democracy

HEC Classbooks and drafts available in English (Several other HEC Classbooks are also currently being translated.)*

Air and Water
Force and Motion (Part 1)
How Many Legs?
If You Could See an Atom
Let's Play with Dry Ice
Objects and their Weight
The Surface of Water

* Currently, most HEC Classbooks and related books are first being published in Japanese by Kasetu-sha (http://www.kasetu.co.jp/). Contact: mail@kasetu.co.jp

The *Kasetsu* Class Album

* Please refer to §2.5, HEC classroom management.

At a primary school,
in the early days of HEC, 1964.

Objects and their Weight

A. Hand out the sheet and read the problem

Once all the students have received the sheet that corresponds to the problem in the HEC Classbook that they will be examining, the class usually starts with having the students read about the problem. The teacher may even demonstrate the setup or the actual experiment (without giving away what the result of the experiment will be).

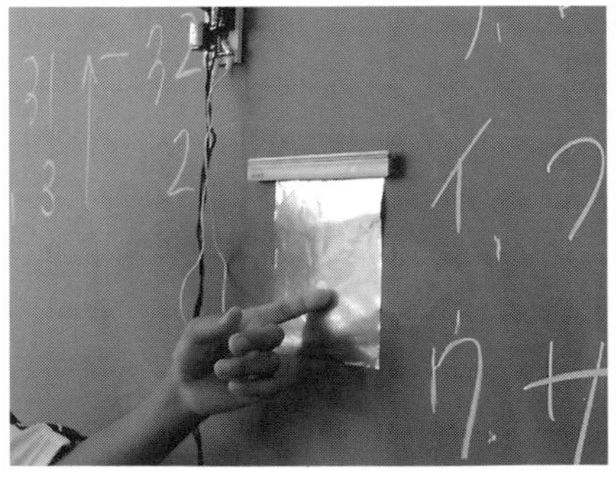

When You Find Free Electrons

'Does the gold origami paper conduct electricity?'

Objects and their Weight

If we float the piece of wood in the water, how will the indicated weight change?

B. Encourage students to come up with an expectation

The teacher may write down choices on the blackboard and may even draw pictures or graphs to clarify the differences between the choices.

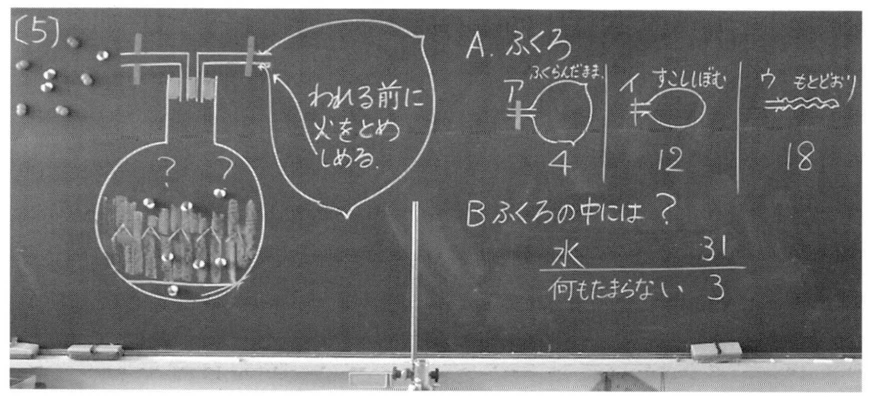

The Three Phases of Matter

After all the students have written down their predictions in their Classbooks, the teacher counts the number of students with each choice and writes the totals on the blackboard.

High school students raising their hands to indicate their expectation.

A science teaching methods class being taught by a university student, simulating a lesson using an HEC Classbook.

C. Let's discuss

A discussion is primarily something that occurs only between people with clear opinions about the problem at hand, but still serves a considerable role for those who are just listening.

D. Changes of expectations

If a student makes a request to change their choice of expectation, the teacher should accept the change and make the necessary corrections to the totals.

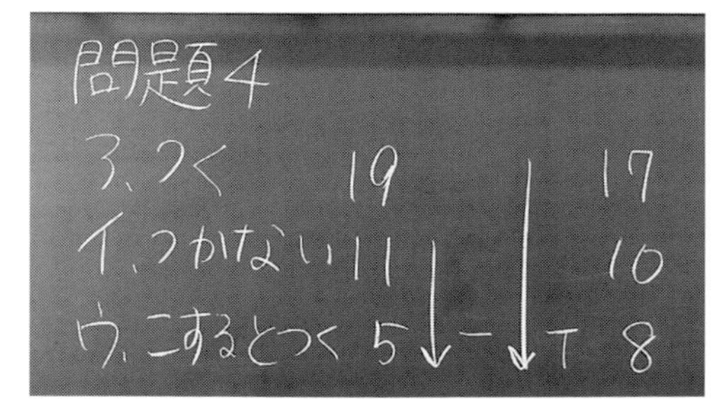

E. Performing the experiment

The teacher usually performs the experiment. Experiments must be performed with only enough precision to clearly determine which expectation is correct.

Will water go up if you suck the 1 m tube? Air and Water

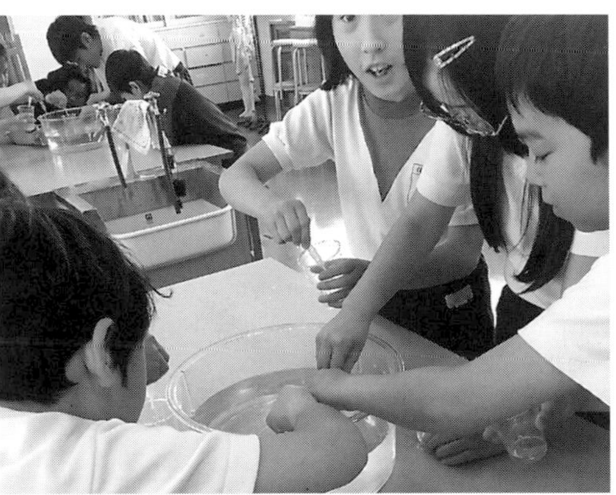

Let's check the results.

They are waiting to see what will happen. Finally, they will know which expectation was right.

They are excited to see the result.

University students waiting to see the results of the free-fall experiment.

The World of Falling Motion

The Three Phases of Matter

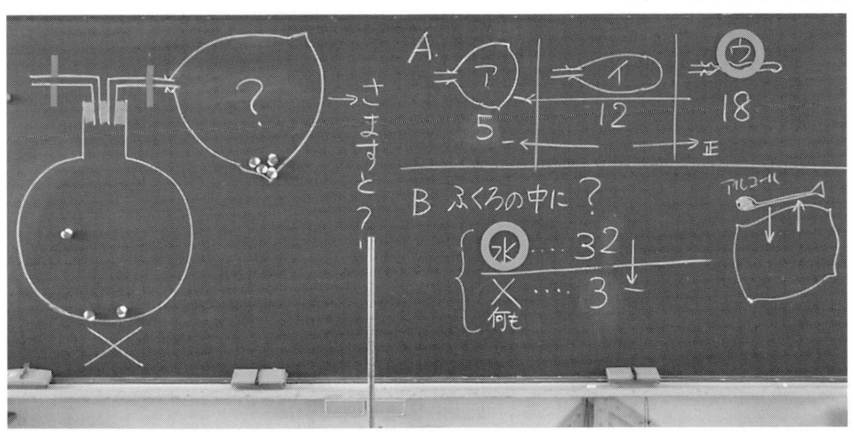

The students should record the results in their Classbook. The teacher must still not explain the result of the experiment. Between problems, there may be passages that explain the experiment result or define certain terms.

F. End-of-Classbook evaluation

Once the class has progressed through every section of the Classbook, students fill out an evaluation of the class, writing down any comments or feelings they have about it.

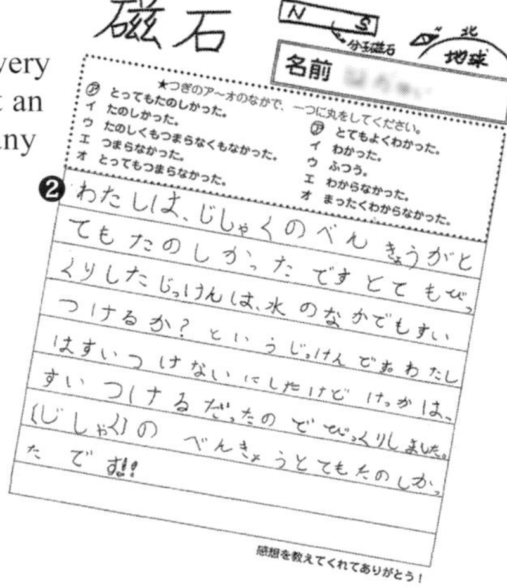
磁石　N S 分子磁石　北 地球

名前

★つぎのア〜オのなかで、一つに丸をしてください。

ア とってもたのしかった。
イ たのしかった。
ウ たのしくもつまらなくもなかった。
エ つまらなかった。
オ とってもつまらなかった。

ア とてもよくわかった。
イ わかった。
ウ ふつう。
エ わからなかった。
オ まったくわからなかった。

❷ わたしは、じしゃくのべんきょうがとてもたのしかったですとてもびっくりしたじっけんは、水のなかでもすいつけるか？というじっけんでわたしはすいつけないにしたけどけっかは、すいつけるだったのでびっくりしました。[じしゃく]のべんきょうとてもたのしかったです!!

感想を教えてくれてありがとう！

❶ 4/13 課題(問題1)の答えが今日分かってすっきりした。

答えが分からないとき　答えが知りたい　答えは何

答えが分かったとき　答えが知れた

We consider the HEC Class a success if most of the students rate the class a 4 or 5. The students' evaluations and comments have helped us delve into research on ways to improve the Classbooks by allowing us to compare across different schools and classrooms.

*The students evaluate the class using a 5-rank scale and write down their comments on the evaluation sheets, as shown in the photos. The 5-rank scale is: 5 = Great, 4 = Good, 3 = So-so, 2 = Boring, 1 = Very boring.

《もしも原子が見えたなら》

年　組　番　氏名

❸ 《もしも原子が見えたなら》の授業はいかがでしたか。初めて知ったことやおどろいたこと，印象に残った先生の話，友達の発言や質問，発言しなかったけれどあのときはこう思った，とくに印象に残った分子や原子など　あなたの感想をおしえてください。疑問に思ったことや，絵をかいてくれてもいいですよ。

⑤ とてもたのしかった
4 たのしかった
3 どちらともいえない
2 あまりたのしくなかった
1 ぜんぜんたのしくなかった

❹ この原子の授業を終えて今までとは驚異的に魅力をもちました。ぼくたち自信もこの世にあるものもすべて原子でできていると知ったときscienceの世界に吸いこまれていくような気分を味わいました。空気中の成分には、ぼくの知らない種類の原子がたくさんあり、ヒトの発展によって公害のもととなる悪い原子が増えてしまうことを知ったので、その対策として、ぼくたちが将来scienceの力で未来を明るくしなくてはいけないと思いました。

年　組　番　氏名

業はいかがでしたか。と，印象に残った先生のったけれどあのときは分子や原子など　あなたに思ったことや，絵をか

⑤ とてもたのしかった!!
4 たのしかった
3 どちらともいえない
2 あまりたのしくなかった
1 ぜんぜんたのしくなかった

❺ 原子の話がたのしかったです。
理科の授ぎょうがある日は、とても楽しみでした。
私が疑問に思った事は、化学者たちは、どうやって、こんなこまかいこと調べたのか、思いました。化学者は、すごいですね!!
もう1つは、「アルゴン」は、何んで「アルゴン」と言う名前なのか思いました。
この勉強をする前までは、「ヘリウム」とか「いおう」を聞いたことがあったけど、どんなのか、知らなかったので、勉強して良かったです。

Notes (Translations of figures in section F)

I felt so relieved that I understood the answer to 'Problem 1' today.

I found the lesson on magnets really enjoyable. The experiment that really surprised me was the one where we looked at whether it was possible to breathe underwater. My guess was that it would not be possible to breathe, but the result was that it is, so I was surprised. I really liked themagnet lesson.

If You Could See an Atom

How was the 'If You Could See an Atom' class? Please tell us your feelings about the things that you learnt for the first time, anything that surprised you, anything that your teacher said that stuck in your mind, your friends' questions and comments, what you think now about things that you did not comment on at the time, and – in particular – atoms and molecules that made an impression on you. We would also like to hear any questions that you might have and see any pictures that you feel like drawing.

Now that this lesson on atoms is over, I feel the most phenomenally astounded that I ever have! When I learned that we and everything else in the world are made up of atoms I felt like I had been drawn into the world of science. Learning that the air is made up of lots of different kinds of atoms that I know nothing about and that the bad atoms, which are a result of pollution caused by human development, are increasing made me feel that we have to use the power of science to counter this and to make the future brighter.

The discussion about atoms was fun. Some days science classes were really fun. What I did have a question about was how chemists have managed to investigate such minute things. Chemists are incredible!! Another thing that I wondered was how did 'argon' come to be called 'argon'? Before these lessons, I had heard of things like 'helium' and 'sulphur' but I had no idea what they were so I'm glad that we learned about them.

Photos courtesy of Masaru Inoue, Shigehiro Kawashima, Aya Nakamura, Noritake Okazaki, Haruhiko Funahashi, Mariko Kobayashi and ASHEC.
Art work by Takanori Hirano.

Appendices

Hypothesis–Experiment Classbook
(*Jugyōsho*)

About the HEC Classbooks Appendices

Representative samples of HEC Classbooks have been included in the Appendices to aid both clear understanding of HEC and implementation in classes.

Objects and their Weight is one of the earliest HEC Classbooks and is an excellent example of the unique features of HEC. This book was also high on Dr. Itakura's list of translation priorities.

Force and Motion was compiled a little later, but concerns research conducted from the very beginning of HEC that reflects Dr. Itakura's innovative ideas for mechanics education. This is one of the HEC Classbooks that is also suitable for use in high school classes.

While *If You Could See an Atom* does not conform to the usual style of HEC Classbooks, it was regarded by Dr. Itakura himself as the most important HEC Classbook. He insisted that atomic theory was absolutely essential for introductory science education, based on his scientific research in this area. He also provided support for classes to help students become familiar with atomic theory through the introduction of molecular models.

The last HEC Classbook I would like to introduce is *How Many Legs?* HEC's approach and ideas are not limited to physics and chemistry. This is an example of an HEC Classbook in biology that enables students to learn in an enjoyable and inspiring way as they make scientific predictions and observations about living things.

There are many other HEC Classbooks that facilitate a style of learning that is at once enjoyable and makes an impression on students. Unfortunately, they have not yet been translated into English.

Dr. Itakura has made great contributions to compiling excellent science books for children as well as conducting specialist research into the history of science. As with his science books, the HEC Classbooks are written with a gentleness that never forces children to learn but instead encourages them to think about things for themselves.

The language of the Classbooks is crucial. It takes considerable time and effort to polish HEC Classbooks in Japanese, and it was very difficult to translate these unique texts. In truth, we are somewhat concerned about how much of his devotion to children, as conveyed in the originals, can be translated into English, so would be very pleased if the message of the original shines through. We would also like to express our gratitude to Ms. Teresa Castelvetere of Trans Pacific Press, who reviewed this book with her in-depth understanding of HEC and its Classbooks.

*

As described in '**2.5 HEC classroom management**,' the teacher will distribute Classbook sheets one by one as the class progresses. The backs of pages with 'Problems' on them have been left blank intentionally so as to avoid students seeing the results of experiments before they have had a chance to formulate their own ideas.

Hypothesis–Experiment Classbook (*Jugyōsho*)

Objects and their Weight

Association for Studies in Hypothesis–Experiment Class

Original Version 1964
English Version 2014

Preface

Hypothesis–Experiment Classes, *Kasetsu* (HEC) centre around the HEC Classbook (*Jugyōsho*).

Objects and their Weight is widely used as one of the most fundamental HEC Classbooks. It is usually used for science classes in primary school, but it can also be used in middle school, high school, or even for adults.

'Weight' in this HEC Classbook refers to the amount of matter in the object, which is equivalent to 'mass' in physics. Though mass and weight are, in fact, different concepts and quantities, we often use the word weight in everyday life to mean mass.

By the age of 10, children generally understand the concept of weight. However, if challenged by a problem they have never experienced before, it becomes clear that they do not always adhere faithfully to principles they have learned.

The aim of this HEC Classbook is to clarify the concept of weight (mass) and its additivity. The concept of weight (mass) is very important to us, because it is the basis for the principle of the conservation of matter, which leads to atomic theory.

Instruments used during experiments

scale, spring scale, beam scale (a larger one that can weigh up to 1 kg, but a smaller one that weighs up to 200 g is also acceptable), 2 graduated cylinders, a beaker or cup

If you plan on doing Enrichment Problems 1 and 2, in Part Two, you will also need a funnel and filter paper, an evaporating dish, a tripod, a wire net, and an alcohol lamp or gas burner.

Materials used during experiments

modelling clay, crackers, wood chips, small rock, sugar cubes, ethyl alcohol, ice, grains of rice, soybeans

sodium bicarbonate and calcium chloride (or saturated salt solution and pure alcohol) for Problem 5 in part Two

Table of contents

Objects and their Weight

Part One: Weight and how it is measured

Question 1

What do you use when you weigh things?

Think of all the instruments for weighing that you know.

You may draw pictures on the other side of this sheet.

Question 2

What units of measurement are used to talk about weight?

Experiment

Would an object's weight remain the same even though a different instrument is used?

Blank page

Problem 1

Have you weighed yourself on a scale?

What would happen to your weight if you tried standing on one leg or crouching?

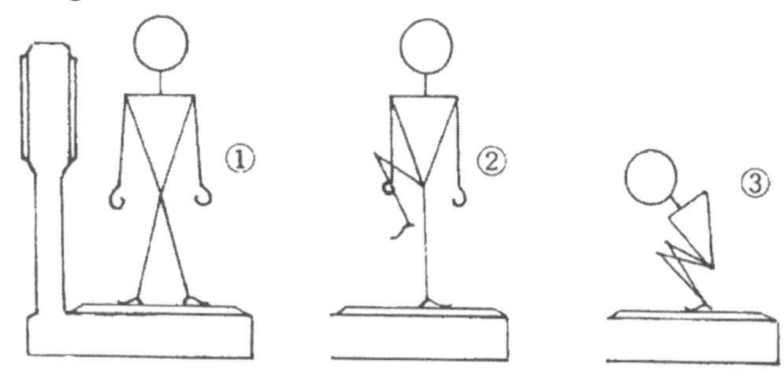

Expectation

Please circle your choice

a. You are heaviest when standing on two legs
b. You are heaviest when standing on one leg.
c. You are heaviest when crouching.
d. They are all the same.

How many people chose each expcctation?

Please share your thoughts. Afterwards, we will try this experiment.

Wait until the scale's needle has stopped before reading the weight.

Results

Blank page

Problem 2

Here is a lump of clay. If we change the shape of the clay to the shapes below and place it on the scale, what will happen to its weight?

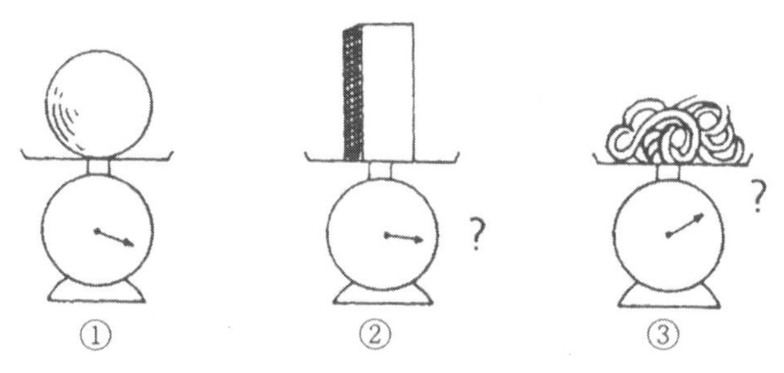

Expectation

a. Shape ① will be the heaviest.
b. Shape ② will be the heaviest.
c. Shape ③ will be the heaviest.
d. Shapes ①, ② and ③ will weigh the same.

Discussion

Why did you choose your expectation? Share your ideas!

Hint: When the shape is changed, do the number of the tiny pieces (atoms) that make up the clay increase or decrease?

Results

Blank page

Problem 3

When a long and slender fish or vegetable is placed on a scale, the ends often hang off of the scale.

A piece of clay has been rolled into a long rod and placed on top of a scale, so its ends hang off.

Will the weight be correctly measured even though both ends of the clay are hanging off of the sides of the scale?

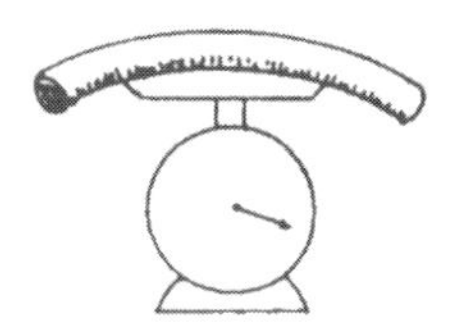

Expectation

a. If both ends hang off of the scale, the weight will be lighter.
b. If both ends hang off of the scale, the weight will be heavier.
c. The weight will be the same even if both ends hang off of the scale.

Please share your thoughts.

Experiment

First, make the clay look like the one in the picture and place it on the scale in the same way. Then, bend the clay in half and check the weight when it is placed properly on the scale.

Results

Blank page

Problem 4

In the last experiment, we discovered that even if the ends of an object are hanging off of the scale, the scale still correctly measures its weight.

Now, what if we gently place one end of the clay on top of something as depicted in the picture? Will its weight still be measured correctly?

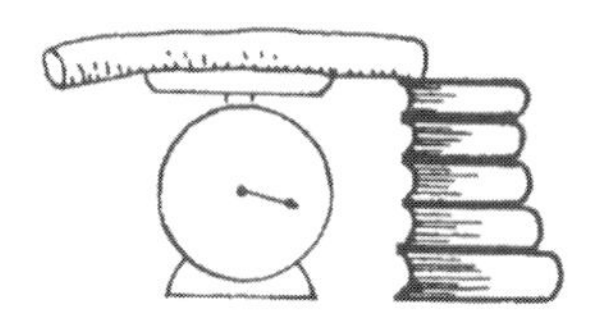

Expectation

a. The scale will correctly measure the weight of the clay, even if one end is placed on top of something.
b. If one end is placed on top of something, the clay will appear lighter on the scale.
c. If one end is placed on top of something, the clay will appear heavier on the scale.

Please share your ideas.

Caution: When using this type of scale, objects should be centred on the scale in order to measure their weight accurately.

Results

Blank page

Problem 5

When crackers were put into a bowl and placed on the scale they weighed _____ g.

If we crush the crackers and place them on the scale again, what will happen to the weight?

Be careful not to drop any pieces when crushing the crackers.

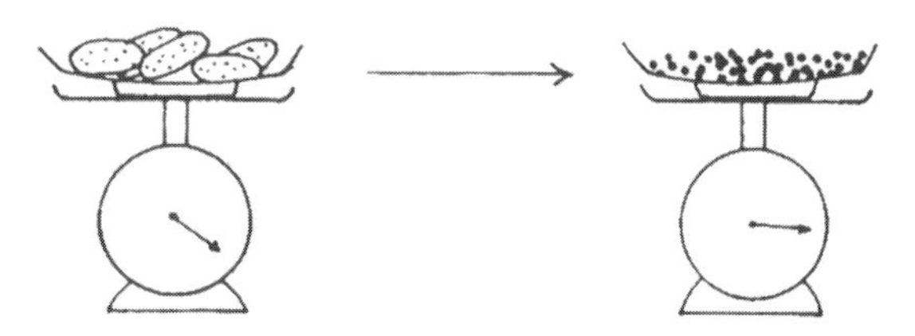

Expectation

a. The cracker crumbs will be lighter.
b. The cracker crumbs will be heavier
c. They will weigh the same.

Discussion

Please discuss your opinions and then do an experiment.

Results

Blank page

Problem 6

When ten sheets of paper were placed on top of each other on a scale, they weighed _____ g.

(1) How much will one sheet of paper weigh?

Expectation

One sheet will weigh [   ] g.

Experiment

After discussing everyone's expectations, place one sheet of paper on the scale and weigh it.

Were your expectations correct?

(2) This one sheet of paper weighed _____ g.

If we cut this paper in half and place it on the scale, how much will it weigh?

Expectation

The half sheet of paper will weigh about [   ] g.

Experiment

After discussing everyone's expectations, cut a sheet of paper in half and weigh it.

Were your expectations correct?

Blank page

Scales and weighing machines

There are many types of scales used to measure weight. We use different scales depending on how large or heavy an object is. The scales we have used for these past experiments are a kind of 'spring scale.' They have a spring below the pedestal of the scale. When we put an object on the pedestal, the extension of the spring corresponds to the weight of the object. There is another type of spring scale that weighs objects hanging down.

It is easy to weigh objects using a spring scale, so it is quite useful. On the other hand, it does not give precise measurements. If it is necessary to weigh something that is less than 1 g, thc spring scale cannot be used. Each line on a spring scale, or graduation, usually represents 10 g or 2 g.

When scientists want to measure an object's weight precisely, they often use a balance beam scale. Your teacher will show you one. There are two plates on a balance beam scale. The object you want to measure is placed on one side, and counterweights are placed on the other side. When the needle is exactly in the middle, the plates are balanced so you can add up the counterweights to find the object's weight. Actually, you do not have to wait for the needle to stop exactly. If the needle swings back and forth *evenly*, the plates are balanced, so you can use that weight.

When measuring an object's weight using a balance beam scale, it is necessary to replace the counterweights many times until the needle is pointing directly in the middle, so it is tedious work. But nowadays there are also scales that show the weight of the object automatically.

Additionally, each type of scale has its own weighing limit, as well as its own minimum unit of measurement, often called 'accuracy.' Sometimes the weighing limit is called 'rated load.' A precision balance beam scale used by scientists can measure down to about 0.001 g (1 mg). An electronic precision balance can measure down to 0.0001g (0.1 mg).

Our spring scale's weighing limit is ______ g and its accuracy ______ g.

The weighing limit of our balance beam scale is ______ g and its accuracy ______ g.

Hands-on Activity:
Weighing different objects using a balance beam

If there is time, we will learn about how to weigh objects using the scales and then weigh various things in the classroom.

How to use a balance beam scale

Place the scale on a flat surface and adjust the needle so that it points to the line in the middle.

1. The object that you want to weigh should be placed on the left plate, and the counterweights on the right plate. This is so that the middle bar will not be in the way when replacing the counterweights with your right hand. (This step is reversed for a left-handed person.)
2. When you want to measure a predetermined weight of medicine or similar object, first place that quantity of counterweights on the left side. You do this because…? Think about why this might be!
3. There is a pair of forceps inside the box with the counterweights. The forceps are not suitable for holding heavy weights, so they should not be used to pick up heavy counterweights as these will most likely drop and be damaged. It is best to use clean, dry hands to pick up the heavy weights. Please use the forceps to pick up smaller weights.

4. If you are not used to using a balance beam scale, it may be difficult to find the correct quantity of counterweights needed to make the needle stop in the middle. It is best to use a heavier weight at first, and if that is too heavy, replace it with the next lightest counterweight. Next, use the smaller counterweights to find the correct spot.
5. When you are finished using the scale, take one plate and put it on top of the other plate. This is done so the arm does not rattle back and forth. If it shakes too much, the edge that supports the plates may get damaged and the scale will no longer be able to correctly measure weight. When two plates are stacked on one side, the arm will not shake back and forth.

A coin weighs about [ ] g.

A postcard weighs about [ ] g.

Problem 7

There is a square drawn on this page that is 1 cm × 1 cm. Please cut out a square this size.

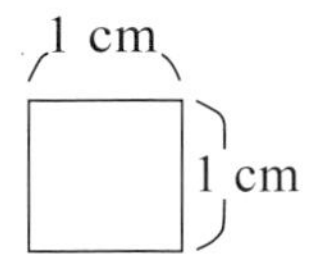

If this small piece of paper is placed on the balance beam scale on your desk, do you think its weight can be measured?

Expectation

a. It can be measured.
b. It cannot be measured.

Discussion

Why did you choose your expectation?

Do you think even this tiny piece of paper has weight?

Experiment

Please place the paper on the scale.

How many grams does it weigh?

Results

Blank page

Question 3

The scale does not move at all when the 1 cm square paper is placed on it. Is there a way to measure the approximate weight of the small piece of paper using this balance beam scale?

Experiment

The next problem, Problem 8, shows us one method of measuring the weight of the small piece of paper. If you came up with a different idea, please try your idea after you have finished Problem 8.

Blank page

Problem 8

When we took ______ sheets of 1 cm × 1 cm paper, cut out the same way as before, and placed them on the scale, they weighed ______ g.

If we know this, how much does one of these pieces weigh?

About ☐ g.

If we cut out a piece of paper that is only 1 mm × 1 mm, how many grams will that weigh?

About ☐ g.

Question 4

Think of something that is extremely heavy and something that is extremely light. Please write down one example of each.

Extremely heavy object

Extremely light object

Discuss the objects you wrote down and compare their weights.

Blank page

Enrichment Problem 1

Very heavy and very light objects cannot simply be placed on a scale and weighed. However, people have come up with various ideas that allow us to measure these objects. Thanks to years of research by scientists, we are now able to weigh atoms that cannot be seen with the naked eye, the distant Moon and Sun, and even our own planet.

Please find out the weight of the objects below by looking through textbooks or asking your teachers.

Object	**Weight**
One hydrogen atom	
One iron atom	
Your weight	
The Moon	
The Earth	
The Sun	

Blank page

Practice Problem 1

Imagine a building block like the one shown in the figure below.

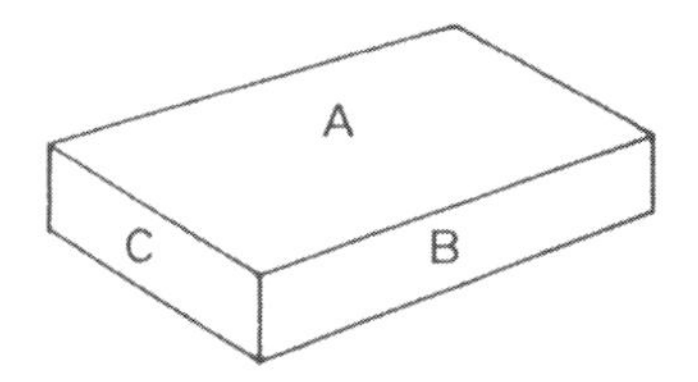

When the block is placed on a scale with its large side (side A) facing down, its weight is 100 g.

If the block is placed on the scale with its smallest side facing down (side C), how many grams will it be?

a. More than 100 g
b. 100 g
c. Less than 100 g

Practice Problem 2

Jill weighs 25 kg. Jack weighs 30 kg. If Jack carries Jill on his back and steps on the scale, how much will they weigh?

a. More than 55 kg
b. 55 kg
c. Less than 55 kg

Blank page

Part Two: An object's change and weight

Problem 1

Here we have a scrap piece of wood. When the piece of wood was placed on a scale, it weighed ____ g.

Next, when a bowl of water was placed on a platform scale, the scale's needle indicated ____ g.

If we leave the bowl of water on the scale and then float the piece of wood in the water, how will the indicated weight change?

Expectation

a. It will increase by the same amount as the piece of wood's weight.
b. It will remain unchanged.
c. It will increase by half of the piece of wood's weight.
d. It will weigh less than before.
e. Other ideas.

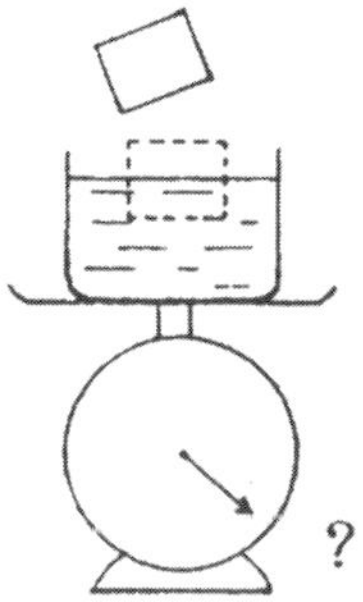

?

Discussion

Why do you think this will happen? Discuss your ideas.

Results

Blank page

Problem 2

When water was placed into a container and then measured on a scale, its total weight was ____ g.

If we place a rock into the water while it is still on the platform scale, how will the indicated weight change?

The rock weighs ____ g.

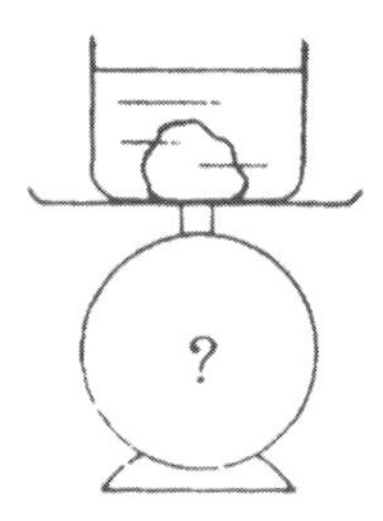

Expectation

a. It will increase by the exact weight of the rock.
b. It will remain unchanged.
c. The weight will increase, but not by as much as the rock's weight.
d. It will be lighter.
e. Other ideas.

Discussion

Why do you think this will happen? Discuss your ideas.

Results

Blank page

Problem 3

First, place a container filled with water and four sugar cubes on one side of a balance beam scale and then place counterweights on the other side until the weight is even on both sides. When you are finished, remove the container and place the four sugar cubes into the water. Stir the water until all of the sugar cubes have dissolved.

Before you place the container back onto the balance beam scale, can you predict how the scale will move now?

Expectations

The side with the dissolved sugar cubes will:

a. become lighter and go up.
b. become heavier and go down.
c. be the same weight and will stay in balance.

Discussion

Why do you think this will happen? Discuss your ideas.

Results

Blank page

Problem 4

A container filled with water and placed on a scale weighs ____ g.

Table salt that was placed separately on a scale weighs____ g.

Take the salt and pour it into the water. Mix it until it is completely dissolved. If we place the salt water back onto the scale, what will the new weight be?

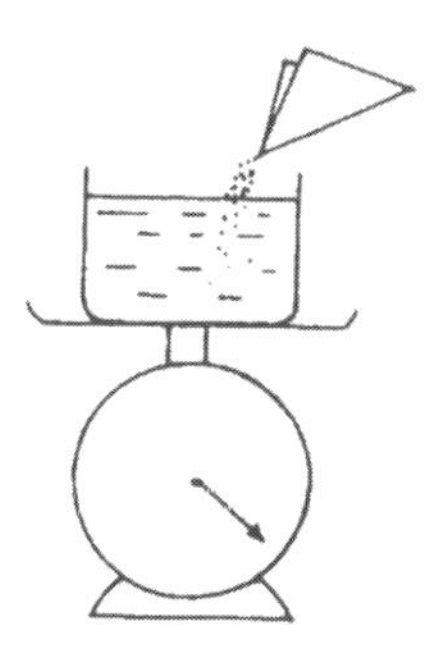

Expectation

About [ ] g.

Results

[ ]

Blank page

Enrichment Problem 1

Take filter paper that is used to filter objects out from liquids, fold it and place it in a funnel. Then, please pour in water that has been made cloudy with chalk powder or something similar and see what happens.

You will be able to see from the result that clean water seeps through the filter paper and drips out the bottom.

When you look at the filter paper, you can see that the chalk powder that was making the water cloudy has now collected there.

Now, place a new filter paper into the funnel and pour in salt water instead. What kind of water will drip from the funnel this time?

Expectations

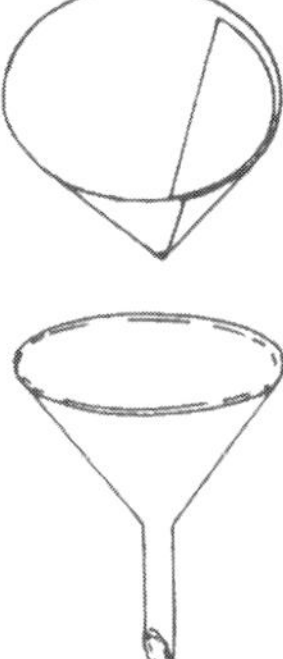

a. The filter will collect the salt, and clean water will drip through.
b. The salt water will pass through the filter unchanged.
c. The water that passes through will be salt water, but less salty.

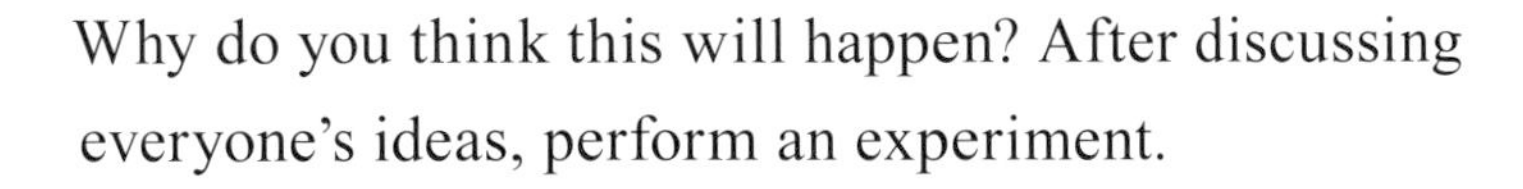

Why do you think this will happen? After discussing everyone's ideas, perform an experiment.

Experiment

Try licking the water that passes through the filter paper.

Results

Blank page

Enrichment Problem 2

Pour a little salt water into an evaporating dish that is resistant to heat. If the bowl is heated from below, as shown in the figure below, what will happen to the salt water?

Expectations

a. Everything will disappear.
b. Only the salt will remain.
c. The water will remain unchanged.

Discussion

Why do you think this will happen? Discuss your ideas.

Results

What have you learned?

Blank page

Problem 5

Here we have two liquids. Both liquids are colourless and clear, but when we mix the two liquids together, a white substance appears and sinks to the bottom. The teacher will demonstrate the reaction of the two liquids so that you know what it looks like.

We use __________ as liquid A and __________ as liquid B.

Now, before mixing the liquids together, we will weigh each liquid separately along with its container. Then we will mix the two liquids and place both containers on the scale.

What will happen to the weight?

Liquid A along with its container weighs _______ g.

Liquid B along with its container weighs _______ g.

Expectation

When we compare the weight of the mixed liquids to the combined weight of _______ g, the mixed liquid will be:

a. Heavier
b. The same
c. Lighter

Discussion and Experiment

After discussing everyone's expectations, perform the experiment.

Results

Blank page

Problem 6 (1)

The materials that are needed for this and the following experiment are 2 graduated cylinders, water, and alcohol.

Pour 50 cm^3 (0.5 decilitres) of alcohol and water into each tube respectively.

If the water is then slowly poured into the alcohol, what do you think will happen to the combined volume of the two liquids?

Expectation (Ignore changes of less than 1 cm

a. It will be exactly 100 cm^3.
b. It will be more than 100 cm^3.
c. It will be less than 100 cm^3.
d. I have no idea.

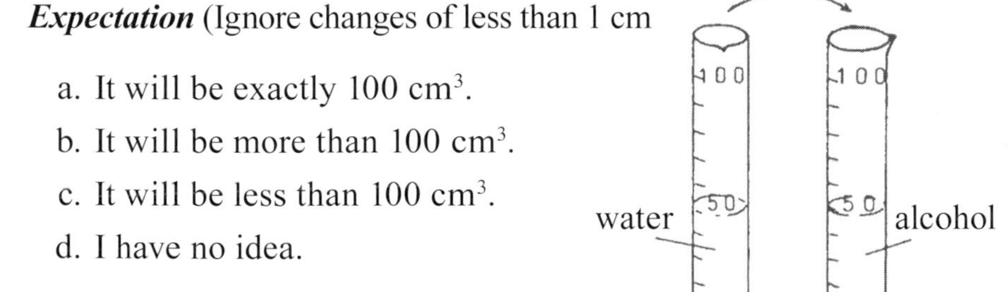

Experiment

After discussing everyone's expectations, perform the experiment and check the result.

Caution: Follow your teacher's instructions on how to properly read the volume on a graduated cylinder. It is a good idea to check the result by comparing the volumes of the two separate mixtures – one is a mixture of two lots of 50 cm^3 water, and the other is the water and alcohol mixed above.

Results

Blank page

Problem 6 (2)

This time measure the weight of 50 cm^3 of alcohol and water.

Alcohol – A volume of 50 cm^3 weighs ________ g.

(________ g including the graduated cylinder)

Water – A volume of 50 cm^3 weighs ________ g.

(________ g including the graduated cylinder)

What will happen to the weight if we mix the water and alcohol?

Expectation

a. It will be the sum of the two liquids. (______ g).
b. It will be lighter than the sum.
c. It will be heavier than the sum.
d. I have no idea.

Discussion

Perform an experiment after discussing your ideas with the class.

Experiment

Make sure you weigh the mixture along with both graduated cylinders. Also, check that the correct volume of liquid is being used.

Results

Blank page

Volume and weight

By now, you should have realised that the combined weight of two objects is always equal to the sum of their individual weights. However, this does not mean that their combined volume will be equal to the sum of their individual volumes.

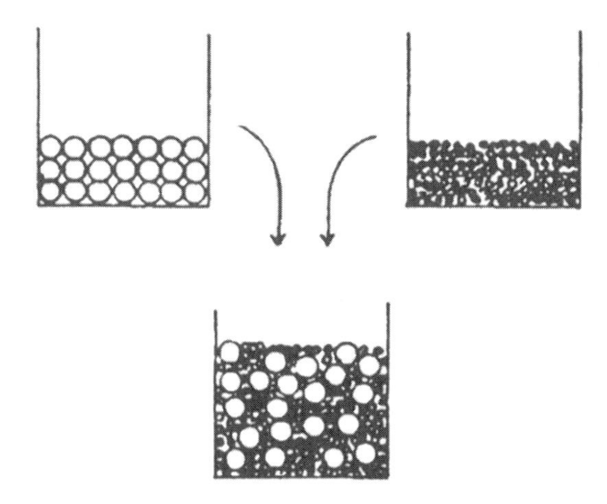

For example, if 1 L of beans and 1 L of rice is mixed together, their combined volume will not equal 2 L. This is because the smaller grains of rice will slip in between the open spaces among the beans. This does not happen only when mixing objects that can be seen by the naked eye. Even when mixing two different kinds of liquids, their combined volume will still be different from the sum of their two original volumes. For example, when mixing 50 cm^3 of alcohol and 50 cm^3 of water, their combined volume does not turn out to be 100 cm^3. It will only be about 97.9 cm^3.

Why does this still happen despite the fact that liquids like alcohol and water look as though they are closely packed? Is it possible that something leaks out without us noticing? That is not what is happening. No, that cannot be it. It must be that there are small gaps between the molecules of water and alcohol and those molecules are slipping in between each other, making the volume less than expected.

However, no matter how many molecules you mix together, causing the combined volume to decrease, the combined weight will never decrease. As long as an object's molecules and atoms do not disappear, the weights of two objects can always be added together to get the combined weight.

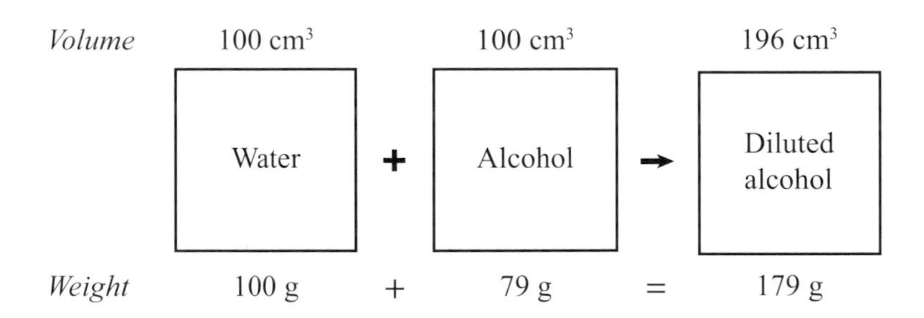

Enrichment Problem 3

First, pour some water into a cup. Then, place an ice cube into the water and measure its weight.

When you are finished, wait until the ice cube has melted and become water. If we weigh it now, what will the new weight be?

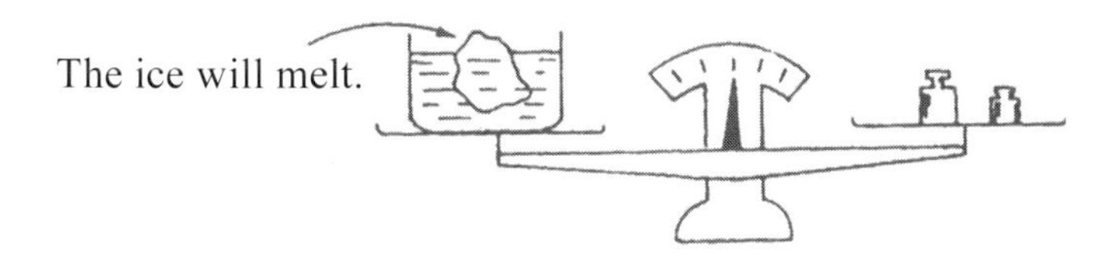

Expectation

a. It will be heavier than before the ice melted.
b. It will be lighter than before the ice melted.
c. It will be the same.

Discussion

Why do you feel this way? Discuss your idea with everyone.

Experiment

Now you try it.

Results

Blank page

Problem 7

A baby is put on a scale, and the scale reads 6,500 g. Right after the baby is weighed, he or she is fed 200 g of milk and then put back on the scale.

How much will the baby weigh after drinking the milk?

Expectation

a. More than 6,700 g.
b. Exactly 6,700 g.
c. Between 6,500 g and 6,700 g.
d. Same as before drinking the milk (6,500 g).

Discussion

Why did you choose your expectation? Discuss your thoughts with everyone.

Experiment

Read the following story about an experiment done by a scientist a long time ago. You probably know which expectation is correct.

For this experiment, you will need a precise scale that can measure to a precision of 100 g. First, weigh yourself on the scale. Then, drink some milk and get back on the scale right away.

Blank page

Santorio Santorio's experiment

The following story is over 350 years old.

There was a professor named Santorio Santorio (1561–1636) who taught at a famous university in Italy called Padua University. He was a medical doctor, but he was very different from other medical doctors of his time. This was because he started to study medical problems quantitatively using many tools. For example, he was the first doctor in the world to use a thermometer to measure a patient's temperature.

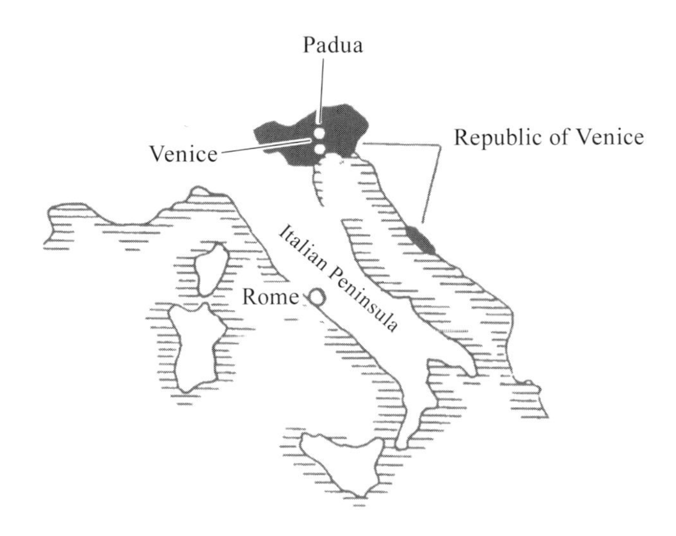

Doctors of that time would check if a patient had a fever or their pulse was fast, as our doctors do today. However, they did not use a thermometer or count the number of heart beats per minute using a watch.

'Well, well, let's check your temperature,' a doctor would say as he placed his hand on a patient's forehead.

'Hmmm, it seems as though you have a fever. It appears slightly lower than yesterday, though.'

In this manner, doctors would decide if patients had a fever, or if the fever had decreased by judging the heat they felt on their hands.

Checking a patient's pulse was done in the same way. A doctor would take a patient's hand and judge if their pulse was fast or slow, based on years of experience. Professor Santorio was not interested in this inaccurate way of doing things. However, there were no thermometers, let alone clinical thermometers, being sold at that time. Scientists still did not have a tool used to measure temperature. So, Professor Santorio went to consult Professor Galileo Galilei who was researching physics also at Padua University.

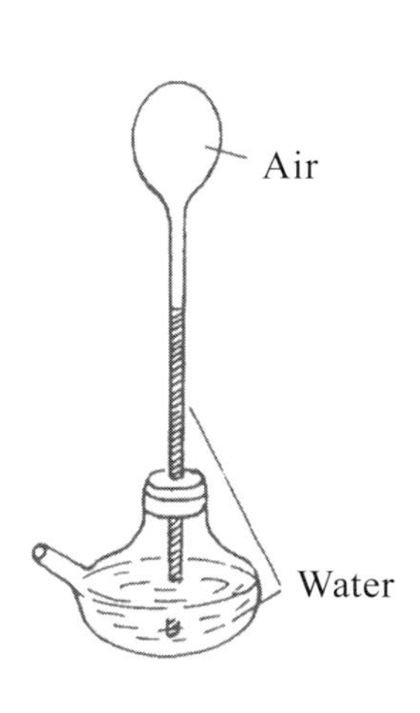

'Galileo, you wouldn't happen to have a device I could use, would you?' When Professor Galilei heard this, he smiled and proudly replied, 'Santorio, my friend. You came at a good time. Actually, I happen to have a tool I invented to measure temperature. I'll perform an experiment and show you.'

Professor Galilei showed him a tool made of glass that looks like the picture on the left.

'There is only air in this round part on the top. And in the bottom, there is only water.' After explaining this, Professor Galilei gently squeezed the ball at the top in order to warm it. The air gradually warmed and expanded, and the expanded air pushed the water down. Professor Santorio was overjoyed.

'I see. By looking at the water level in the pipe, you can determine the temperature.'

Professor Santorio quickly returned to his room and improved the machine so that it could measure a human's body temperature. He made the ball smaller so that it could fit in a patient's mouth, made the pipe much longer and twisted, and marked divisions on it so the temperature could be read. When compared to today's thermometers, this one is very inconvenient. However, it is the world's first clinical thermometer.

Professor Santorio also made an instrument that could measure a person's pulse. By using Professor Galilei's research once again, he invented a machine that could measure a person's pulse using a pendulum.

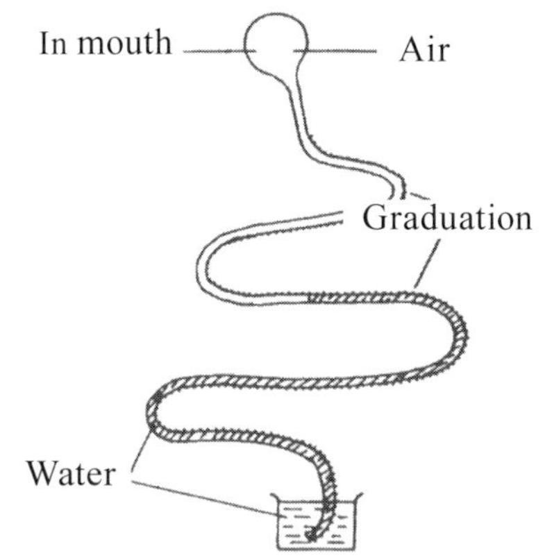

While these inventions were great, his most interesting research is most likely an experiment he performed using a scale. In his book, he writes, 'I have spent most of the last 30 years of my life on a scale.' Professor Santorio built a very large beam scale and on one side of the scale he hung a chair that he would always sit in. Whether he was eating or working, he always sat in the chair. In so doing, he was able to examine how his weight changed at all times. He was able to discover many interesting things by checking his weight every day. For example, he was happy to discover that 'when one's weight suddenly increases or decreases, they are in poor health.'

Of course, he also examined how his weight changed before and after eating. He would first weigh his dinner and then be sure to eat all of it. He found that his weight would always increase by exactly the amount the food weighed. In addition, he also examined how his weight changed while working, resting, and even sleeping. He then proceeded to write a book called *De Statica Medicina* (*On Medical Measurement*) that gave detailed accounts of all his research. This is an excerpt taken from his book.

> I have spent the entire day sitting on my scale and examining the changes in my weight, but I have seen very minimal changes in my weight compared to the previous day. However, during this time I ingested 3,600 g of food and the sum total of my excreta was only 1,350 g. This means that 3,600 − 1,350 = 2,250 g were expelled from my body without my knowledge.

So, where did the 2,250 g go? Professor Santorio thought that maybe the moisture that escapes in a person's breath might be the cause and proceeded to test this theory. He placed a glass bottle over his mouth and exhaled into it. He weighed the moisture that collected in the bottle and was able to predict that humans lose about 250 g of moisture a day in their breath. Professor Santorio thought to himself, ‘This means that something else is escaping from my body without me noticing it.’

He started searching for this mystery cause, and after trying various experiments he made an amazing discovery. He discovered that all humans continuously evaporate water from their entire bodies without realising it. This sweat is also expelled from the body while sleeping. Furthermore, humans excrete twice as much sweat while asleep than they do while awake. He found that more than half of all the food and liquids that we consume in a day is expelled into the air as sweat without us even noticing it.

After verifying his results, Professor Santorio came to a realisation. He thought, ‘When a person’s weight suddenly increases, it causes them to be sick. This must be because their body is not releasing enough water by sweating!’ From that point on, whenever a patient would see Professor Santorio, he would feed them a decided amount of food and weigh their excrements. If he found that the patient was not sweating enough, he had them drink medicine that would cause them to sweat more.

During Professor Santorio’s time, there was not much scientific research being done in the field of medicine. Therefore, there were many times when he drew false conclusions. On the other hand, having the forethought to use scales and thermometers in his research was amazing for his time. Thanks to scientists such as Professor Santorio, medical research from that point onward started to advance more scientifically.

Blank page

To contact the Association, please send emails to
mail@kasetu.co.jp

Translators
Daniel Charles Whiteman, Alexander Clemmens,
Takahiro Kinoshita, Mariko Kobayashi

Translation advisers
Haruhiko Funahashi, Nobuo Takahashi

Editors
Mariko Kobayashi, Haruhiko Funahashi

Publisher
Kasetu-sha http://www.kasetu.co.jp/

Hypothesis–Experiment Classbook (*Jugyōsho*)

Force and Motion
Part One: Force and Acceleration

Association for Studies in Hypothesis–Experiment Class

Original Version 1971
(Revised in 2007)
English Version 2018

The aim of 'Force and Motion'

Kasetsu Hypothesis–Experiment Classes (HEC) centre on *Jugyōsho*, the HEC Classbooks. This HEC Classbook is designed for use in middle school, but it may also be of use in both elementary school (grades 5 and 6) and in high school. We assume that students are already familiar with representing forces as arrows and with the basic idea of the equilibrium condition of forces.

Traditional education in dynamics is frequently concerned only with calculations and solving problems, ignoring any connection to real life. We therefore developed this Classbook with a focus on problems, expectations, discussion, and experiments; no emphasis is placed on mathematical calculation.

There should be no difficulties in conducting the experiments in this HEC Classbook, and all calculations are straightforward. It is our belief that any teacher using this Classbook will be able to ensure that no child is left behind.

Table of contents

Part One: Force and Acceleration

Force and Motion

Part One: Force and Acceleration

Problem 1

Here is a block weighing ____ g.

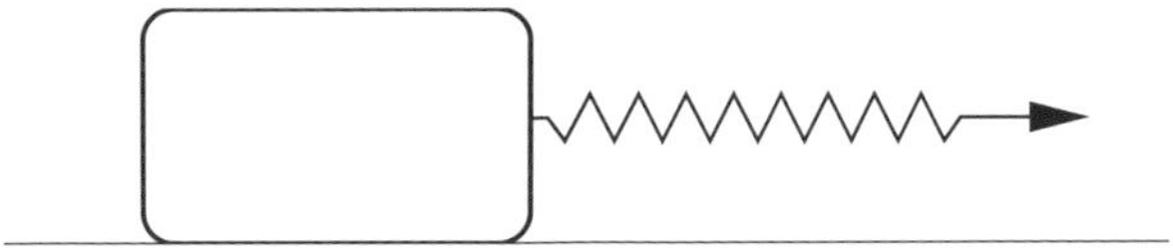

How many gram-force (gf) do you need to move the block horizontally on a smooth desk?

Expectation

a. It will take a force of exactly _____ gf.
b. It will take a force greater than _____ gf.
c. It will take a force less than _____ gf.

Discussion

Why do you think so? Share your thoughts, then do the experiment.

Results

Blank page

Problem 2

Now, put a series of round pencils under the block from Problem 1, and then try pulling it horizontally as before.

Compared to the previous experiment, how much force is required?

Expectation

It will take

a. about the same as before.
b. about half as much as before.
c. about one tenth as much as before.
d. much less than one tenth.

Discussion

Why do you think so? Share your thoughts, then do the experiment.

Results

Blank page

Frictional force

To lift up an object that weighs 1 kg, it takes more than 1 kilogram-force (kgf). However, this object can usually be moved horizontally by less than 1 kgf. The Earth pulls down on a 1 kg object with a force of 1 kgf. At the same time, this force is in balance with a force from the surface that the object pushes against.

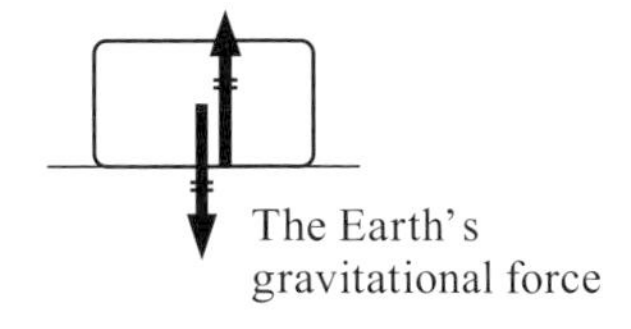
The force from the surface that pushes back against the object.
The Earth's gravitational force

When an object moves across a surface, the object and the surface will rub against each other. This force, which works to stop the object's motion, is called the **frictional force**, or just **friction**. Therefore, we will only be able to move the object if the total force acting on it is greater than the frictional force produced when it rubs against the surface. The magnitude of the frictional force depends on the nature of the surfaces of the objects which rub together. In general, the magnitude of the frictional force is between about one-tenth and half of the gravitational force acting on the object.

However, if we use something like a roller or a wagon, the magnitude of the frictional force will be several hundred times less because the original object and the surface will hardly rub together at all.

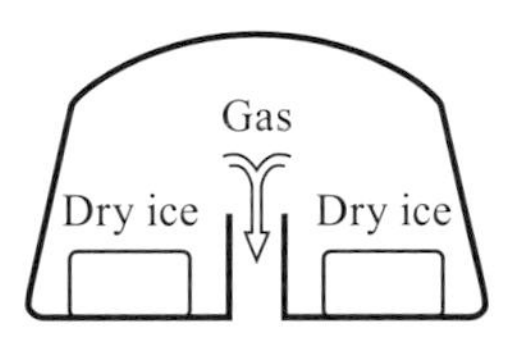

If the gas comes out here, the object will not rub against the surface of the desk directly, so the magnitude of the frictional force is almost zero.

Recently, many devices like the one in the figure above have been developed which use a cushion of air to decrease the frictional force to almost zero. The magnitude of the frictional force will be greater before the object moves but will decrease once it starts moving. Accordingly, the frictional force when movement begins is called **maximum static friction** and when the object is in motion, the frictional force is called **kinetic friction**. This is how we distinguish one from the other. Using these new terms, we can say that, in general, kinetic friction is __________ than maximum static friction.

Problem 3

Here is a wagon.

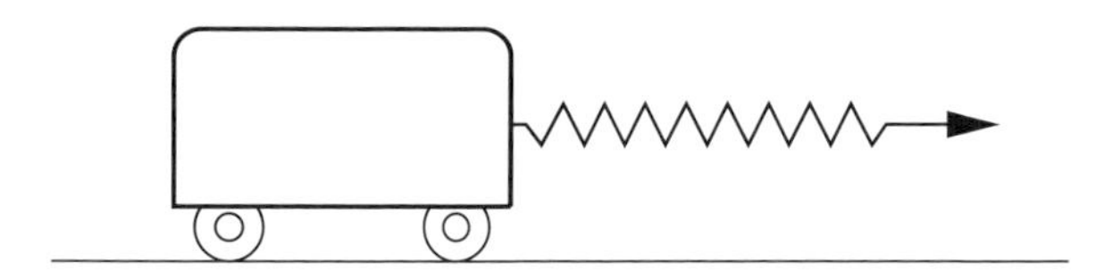

Keep pulling, making sure that the extension of the spring does not change.

If we continue pulling with a constant force, as in the diagram, what do you think will happen to the motion of the wagon?

Expectation

a. It will always move at the same speed.
b. The speed will gradually increase, then become constant.
c. It will get faster and faster.

Discussion

Why do you think so? Share your thoughts, then do the experiment.

Experiment

Extend the spring to a length of _____ cm while holding the wagon to prevent it from moving, then release the wagon and continue pulling, keeping the spring the same length.

Results

Blank page

Problem 4

The reason why an object falls is that it is pulled down by the Earth's gravitational force.

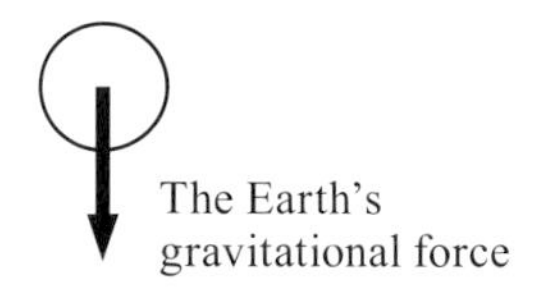
The Earth continues pulling the object down while the object is falling.
The Earth's gravitational force

So, if we drop a marble from a height of 1–2 m what do you think will happen to its falling speed?

Expectation

a. The marble will fall at a constant speed.
b. The speed will gradually increase and quickly become constant.
c. The speed of the marble will continue to increase as it falls.

Discussion

Why do you think so? Share your ideas.

Experiment

How can we find out the speed? Share your thoughts, then do an experiment.

Results

Blank page

Force and acceleration

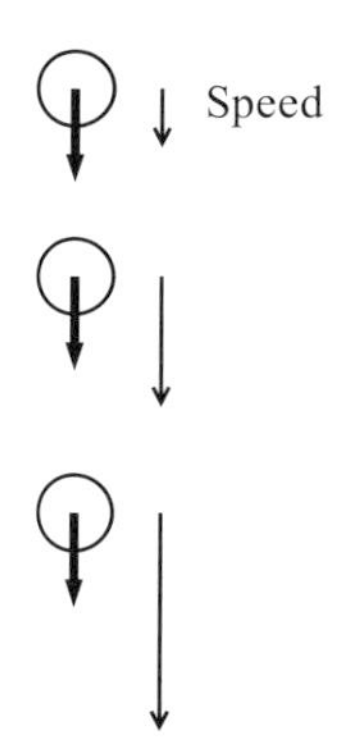

The speed of a wagon rapidly increases when pulled by a constant force.

In the same way, a marble dropped from a height will rapidly increase in speed because it is being pulled down by Earth's gravity, which is constant. The speed of the wagon and the marble can only increase up to a certain limit, because the air prevents them from gaining any more speed. If there were no air to resist their movement, the speed of the marble and the wagon could increase without limit.

By examining the speed of the marble, we find that it falls at a speed of 1 m/s (to be more exact, 0.98 m/s) 0.1 seconds after it starts falling. It falls at a speed of 2 m/s after 0.2 seconds and at a speed of 3 m/s after 0.3 seconds. The falling speed increases with time. The speed reaches 10 m/s in 1 second and 20 m/s in 2 seconds.

Why does the speed of the object increase if you keep applying force? Here are some possible answers.

When you apply force to an object at rest, it starts to move. What will happen if you apply constant force after it has started moving? Because of the extra force, the object moves faster than before.

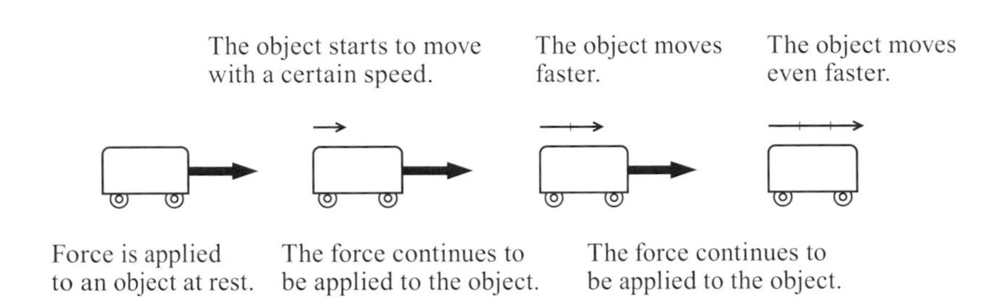
The object starts to move with a certain speed.
The object moves faster.
The object moves even faster.
Force is applied to an object at rest.
The force continues to be applied to the object.
The force continues to be applied to the object.

So, if we keep applying force, the speed of the object will get faster and faster. In this way, if we keep applying the same force, the speed will keep increasing. When force is applied to an object, its speed changes depending on how much force is applied and for how long. When we study the movement of an object, it is important to consider the quantity force × time. Physicists call this **impulse**. When considering force and motion, not only the speed but also 'how the speed increases' is important. The word **acceleration** is used to describe how the speed increases after adding force, as well as how much the speed increases within a fixed interval (1 second).

Problem 5

We can draw a graph to show the speed of the falling marble, as below.

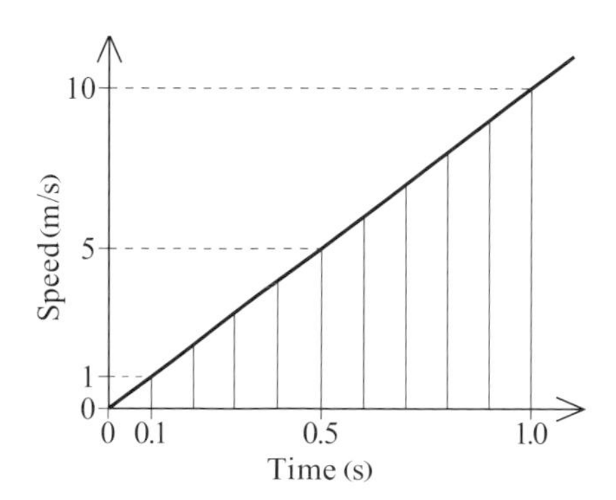

Try to estimate how many metres the marble falls within the first second.

Expectation

I think it falls roughly [ ] m.

Discussion

Why do you think so? Share your thoughts.

Results

Make a note of the results shown to you by your teacher.

Blank page

Constant acceleration

In the case of the falling marble, the speed increases the same amount in the same time (for instance, every 0.1 seconds), so we call this **constant acceleration**. In the same way, if you continuously pull a wagon with a constant force, as in Problem 3, the wagon moves with constant acceleration.

When you have constant acceleration the graph of speed against time is always a straight line as shown below.

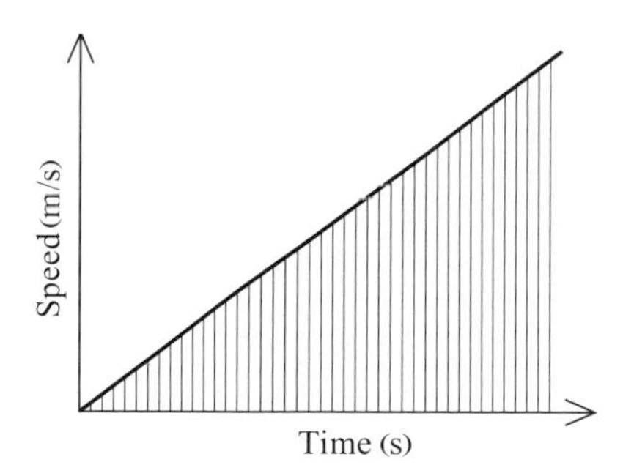

On the other hand, if the speed does not change at all, then we have so-called **uniform motion**. In this case, the graph of speed against time is a horizontal line, as shown below.

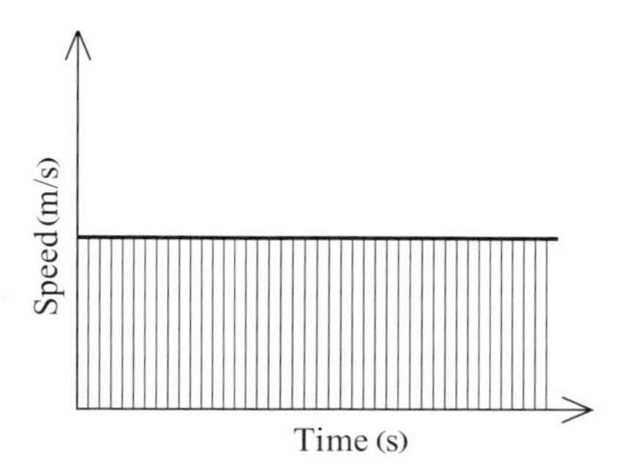

How to calculate distance from the speed–time graph

We can calculate the distance travelled in a fixed time by using this formula: speed × time. In the case of uniform motion, the speed is constant, so we can calculate the distance by multiplying the speed by the time.

But, how can we calculate the distance travelled if the speed changes with time?

For example, the graph below shows the speed of a train between two stops. It is possible to use this graph to calculate the distance between the two stops if we first work out the average speed then multiply it by the time.

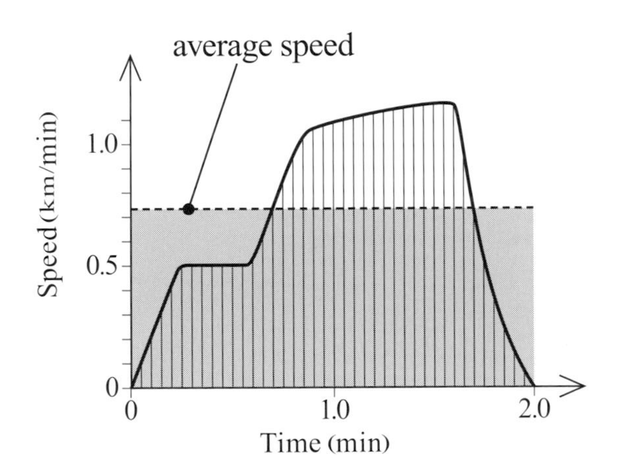

It is simple to work out the average speed for constant acceleration: it is equal to the speed at t_1, which is exactly halfway between t_0 and t_2, as in the graph below.

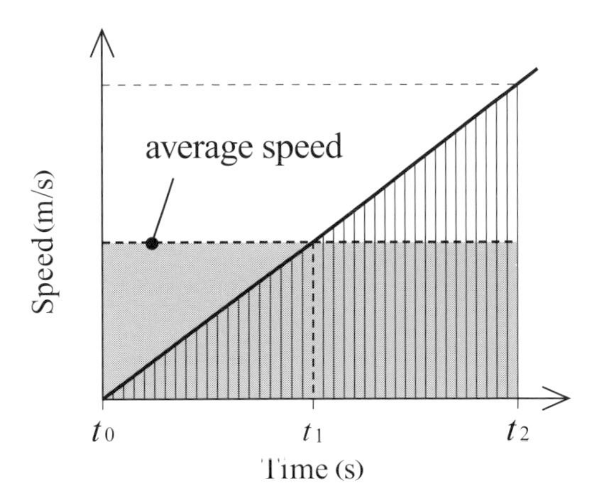

Therefore, in Problem 5, you can calculate the distance that the marble falls in 1 second as follows: 5 m/s × 1 s = 5 m, where 5 m/s is the marble's average speed.

Blank page

Problem 6

The speed of a falling marble changes as in the graph. Using the graph, please answer the following three questions.

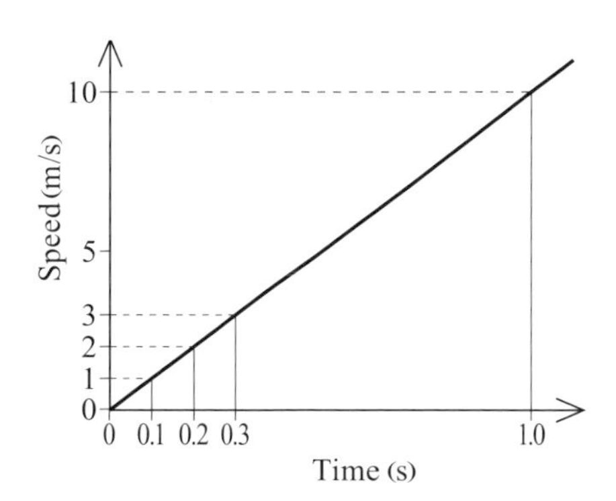

1. How far does the marble fall in 0.1 seconds?

   *The average speed between zero and 0.1 seconds is ☐ m/s.

   *The distance travelled in 0.1 seconds is

   ☐ m/s × ☐ s = ☐ m
   = ☐ cm

2. How far does the marble fall in 0.2 seconds?

   ☐ m/s × ☐ s = ☐ m
   = ☐ cm

3. How far does the marble fall in 0.3 seconds?

   ☐ m/s × ☐ s = ☐ m
   = ☐ cm

Blank page

Problem 7

Putting together our answers from Problem 6 (parts 1 and 2), we get the following:

*The marble falls ______ cm in the first 0.1-second interval.

*The marble falls ______ cm in the next 0.1-second interval.

The speed over the latter interval is greater than in the first interval. Therefore, the distance travelled in the latter interval is ______ times as long as in the first interval even though both intervals are 0.1 seconds long.

Now, how much further does the marble travel in the third 0.1-second interval compared to the first?

Expectation

The distance that the marble falls in the interval between 0.2 and 0.3 seconds is [ ] times as long as the 5cm that it falls in the first 0.1-second interval.

Calculation

Share your ideas and then check by calculating the distance.

Blank page

Problem 8

How much further does the marble fall in the next 0.1-second interval (from 0.3 s to 0.4 s) compared with the first 0.1 seconds (5 cm)?

☐ times.

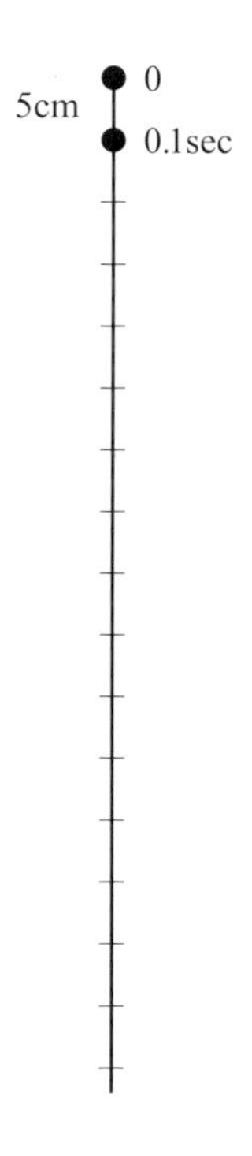

The line above is marked off every 5 cm.

Indicate on the diagram the distance travelled by the falling marble after 0.1, 0.2, 0.3 and 0.4 seconds.

Blank page

Problem 9

A marble rolls down a long slope, as shown in the diagram. How will its speed change over time?

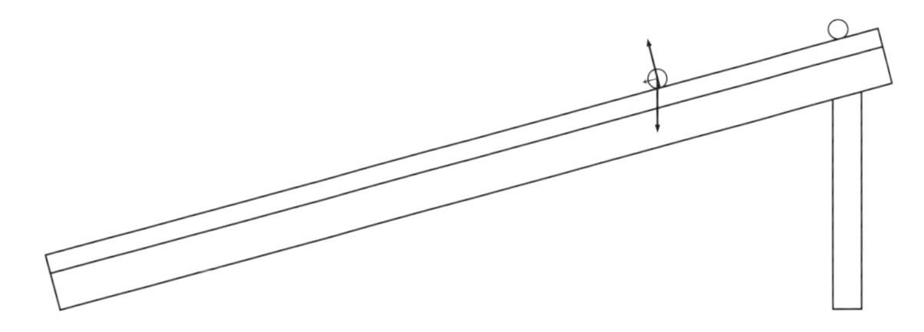

Expectation

a. The speed of the rolling marble will be constant.
b. It will increase initially then eventually become constant.
c. It will keep increasing.

Hint: The marble on the slope always experiences forces as shown in the diagram.

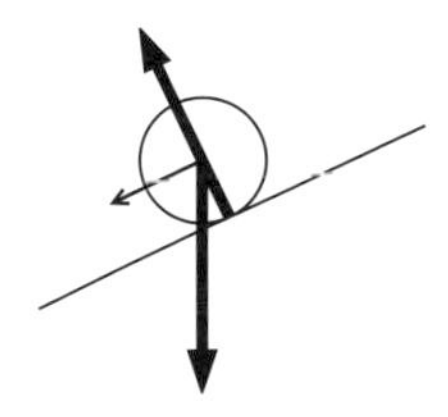

Discussion

Why does it behave like this? Share your thoughts, then experiment using a curtain rail as the slope.

Results

Blank page

Enrichment Problem

Use the same set up as before, but this time with two marbles. First place one marble at the top of the slope. When it rolls 10 cm down the slope, place another marble at the top of the slope and release it as before. What can we say about the distance between the two marbles?

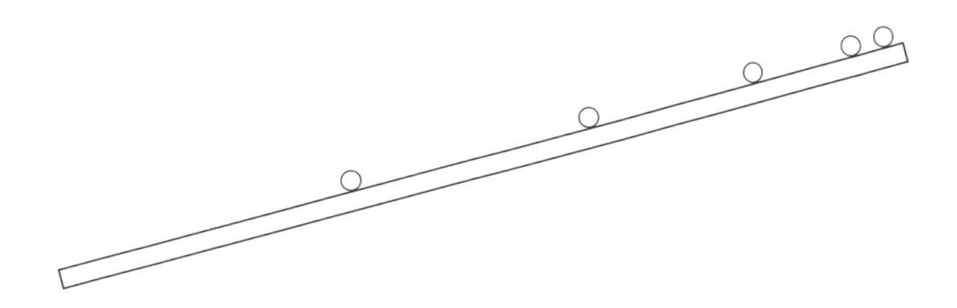

Expectation

a. They will remain the same distance apart.
b. They will get closer together.
c. They will get further apart.

Discussion

Why do they behave like this? Share your thoughts, then do the experiment.

Results

Blank page

To contact the Association, please send an email to mail@kasetu.co.jp.

Translators
Mariko Kobayashi, Gianni Giosue,
Tomoko Kitamura, Tomoko Hasegawa,
Junichi Arakawa

Translation advisers
Dorian Lidell, Hiroshi Yokotani,
Haruhiko Funahashi

Collaborators
Toshihide Yamaji, Takanori Hirano

Editors
Mariko Kobayashi, Haruhiko Funahashi

Publisher
Kasetu-sha http://www.kasetu.co.jp/

Hypothesis–Experiment Classbook (*Jugyōsho*)

If You Could See an Atom

Association for Studies in Hypothesis–Experiment Class

Original Version 1976
English Version 2010

The Educational Significance of 'If You Could See an Atom'

Kiyonobu Itakura

'If You Could See an Atom' is the most well-established Hypothesis–Experiment Classbook (HEC Classbook). Originally this HEC Classbook was meant to be a science picture book – one of *Dr. Itazura's science books*[*] – intended to introduce the basic concepts of atomic theory to 10–12-year-old children.

However, the contents of this book were welcomed not only by children, but also by school teachers who used it as a teaching plan for classes in middle and senior high schools. Finally, this plan was successfully used even in 6–8-year-old children's classes. These results have exceeded the author's expectations.

When traditional educators hear that we use atom models in our classes, they regularly claim that elementary school students cannot possibly understand or accept such a difficult concept. But, actually, the students have completely welcomed the contents of this HEC Classbook without any concern for these educators' misgivings. Why do children like lessons that use atomic models? This Classbook first introduces the fact that 'All the things that exist in this universe are made of atoms.' Then, children begin to ask questions such as, 'Is my body also made of atoms?' 'How about this desk?' 'How about this book and cup?' 'How about the Sun or the stars?' When the teacher tells them 'Yes, everything around you is made of atoms, exactly,' the students say, 'Really!?' It is not because they do not trust the teacher but because they are amazed by the discovery.

---

[*] *Itazura*, which resembles Itakura, means mischievous or funny in Japanese.

Children tend to trust the results of scientists' work. Modern science education should keep this in mind and be careful not to betray children's trust when teaching them what they want to know. It is with this in mind that we promote the use of 'If You Could See an Atom,' which has been welcomed by children and also by their parents.

In a typical HEC Classbook, we provide interesting and challenging problems with experiments and let every student guess and discuss what the results might be before performing the experiment. The children are deeply motivated by the fact that they can discover the problem by themselves. This is also one of the major goals of HEC. Children will lose the sense of pleasure and challenge in investigating the problem by themselves if they already know the result. This is why we have decided that the problems in the HEC Classbook should not be released as a matter of course.

'If You Could See an Atom' is exceptional among HEC Classbooks, because it does not contain questions: it only has statements and drawings, or models of atoms and molecules, which are based on scientists' research. This particular Classbook is more useful if taught in a class. Children can use the atom models in different ways: such as, looking at atom or molecule models which the teacher shows them, making the models by themselves, or discussing atoms and molecules. They acquire knowledge of atoms and molecules through group studies in their class. The learning environment is very relaxed and enjoyable.

The 'If You Could See an Atom' Classbook has also succeeded in letting students imagine the concept of molecular motion in their mind. Of course, in real life, gas molecules such as oxygen, nitrogen and water [vapour] are moving around very actively all together. Moreover, nowadays, students can see computer simulations of molecular motion. Many people know that gas molecules are flying around in a vacuum, but they often think that the motion of the molecules might be decreased by air resistance. The so-called air resistance does not affect the motion of gas molecules in a vacuum. The only factor affecting motion is the collision of molecules in the air.

Most of the children who learned using this Classbook became familiar with atoms and molecules. But many others, usually high school or university students, often express distaste for the topic of atoms and molecules and say, ‘We will never understand those sorts of things.’ It is very important for us to be familiar with an easy and understandable model of atoms from our childhood. We hope that this plan will be a good way of sharing the magnificent experience of thinking scientifically with atom models.

The sizes of the atom and molecule models in this book are magnified one hundred million times. The original size of atoms and molecules is based on the van der Waals radius. It was named after a Dutch scientist, Johannes Diderik van der Waals (1837–1923), who researched the size of gas molecules. The models in this book are called Stuart-type models after the German scientist, Herbert Arthur Stuart (1899–1974). He proposed making molecule models based on the van der Waals radius in 1934, and after that they came into practical use. The Stuart-type model is also called the solid volume model because the real size of atoms is used as a reference. There are various types of models which are used by scientists for different purposes. Recently the solid volume model has become the most popular. The Stuart type model has enabled us to imagine real atoms and molecules.

If You Could See an Atom

Atoms and molecules in the air

Have you ever seen the air?

Have you ever seen an atom or a molecule?

Have you heard the word atom?

Atom means 'inseparable' in Greek.

How do you say 'atom' in any other languages that you know?

Have you ever seen an atom?

There are atoms all around us.

The stones and soil are made of atoms. Paper and trees consist of many atoms.

Iron and glass, air and clouds – they are also made of atoms.

Of course, our bodies are made up of a huge number of atoms.

All things around us are made of atoms even though we cannot see them. An atom cannot be seen by the naked eye just like you cannot see the air.

Have you ever seen planet Earth?

Everyone has seen the ground, but we cannot see the Earth as a globe. (We have recently been able to see that the Earth is round from photographs from space.)

No one can see the entire planet, but we know the Earth is round. We can imagine it even though we cannot look at the entire Earth directly.

When we are at the seashore, the horizon appears to be a straight line; it does not look curved.

But we can draw a picture of the Earth and make a model of it.

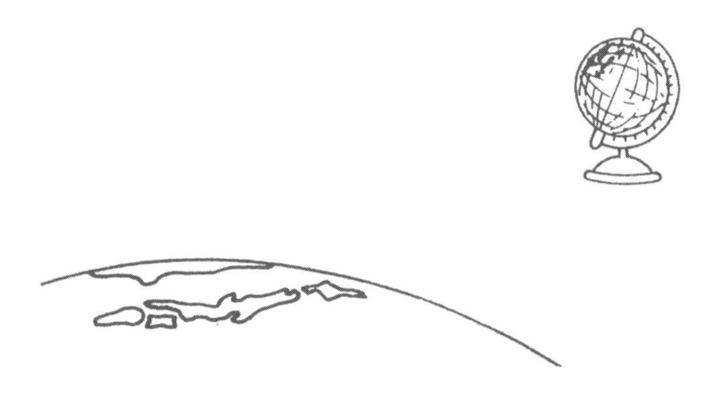

The Earth is too large to be seen all at once, while the atom is too small to be seen with the eye.

However, scientists know the atom very well. They have researched it in detail by:

- exercising imagination,
- carrying out difficult calculations,
- and doing various experiments.

Scientists can draw pictures of the atom and make models as if they had seen atoms with their own eyes. They can also take an electron microscopic picture of atoms today.

Draw a picture of an atom and make a model with your friends.

Question 1

If you were able to see the world of atoms, what would the air around you look like?

Imagine what it would look like and draw a picture.

Blank page

This is a picture of the air drawn by scientists as they imagine it.

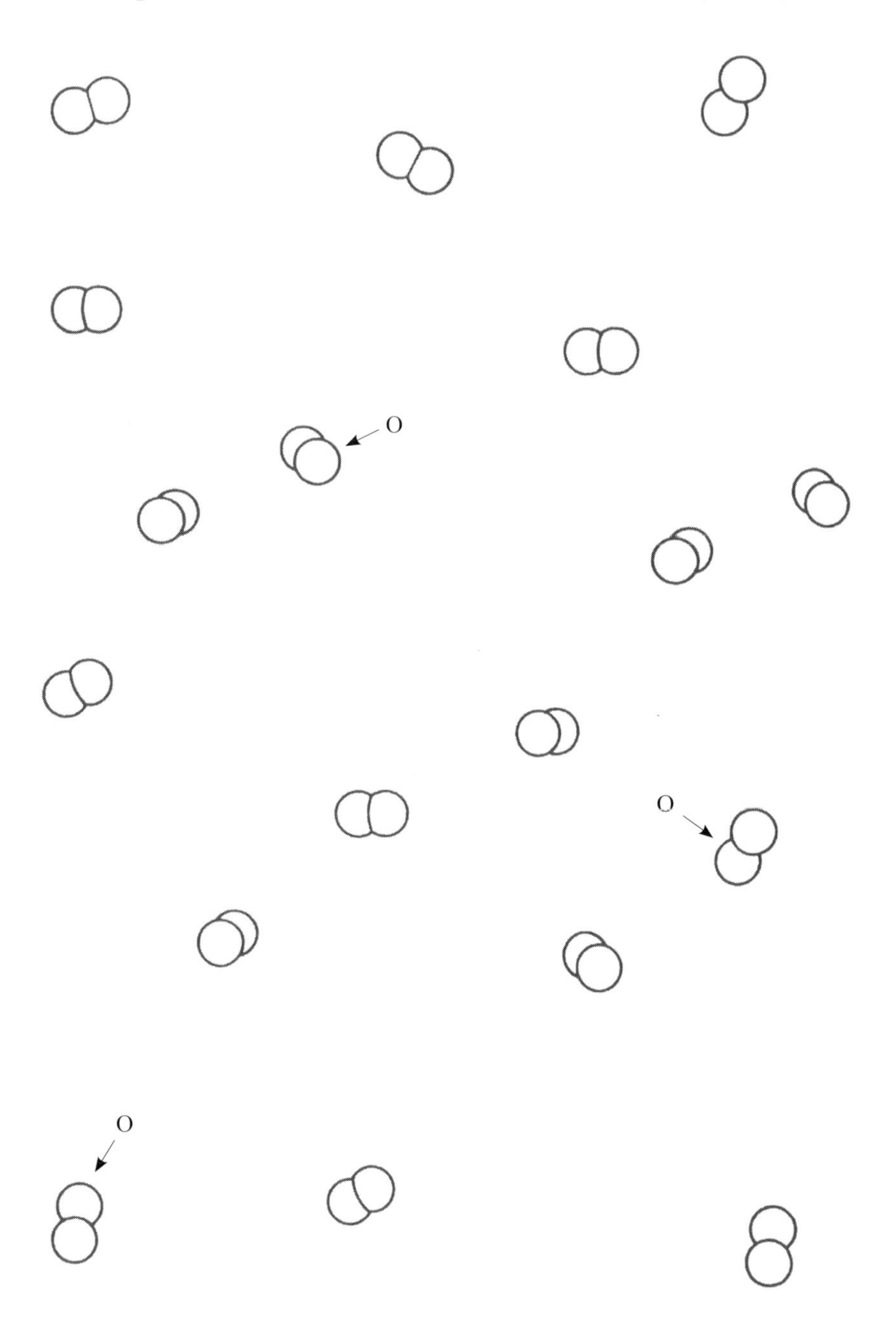

In this picture you can see some small particles. These particles are flying about actively. They sometimes collide and are sent bounding away.

The place between particles is empty. There is nothing whatsoever there. It is completely empty. Scientists call this true emptiness a 'vacuum.'

The particles of air run through the vacuum at a very high speed. They move in a straight line until they hit each other because this space is empty.

There are two kinds of air particles. Imagine that you picked them up and enlarged them. (See the next page.)

In both of these particles the two round shapes are firmly intertwined and bonded.

Each of these individual round shapes is an 'atom.'

Two *oxygen atoms* have the ability to combine firmly.

Two *nitrogen atoms* also have the ability to combine firmly. The two atoms are bonded firmly in combination, and do not part from each other easily.

Scientists call the two oxygen atoms bonded together an oxygen molecule.

The nitrogen molecule has two nitrogen atoms that are bonded together.

Your teacher will show you a model of the oxygen atom and the nitrogen atom. Look at the molecule model made by two atoms. If there are atom models available, try to make a molecule with them by yourself.

Look at the models below and colour them.

Models of atoms are coloured but, in reality, atoms do not have any colours. We colour them to know which is which easily.

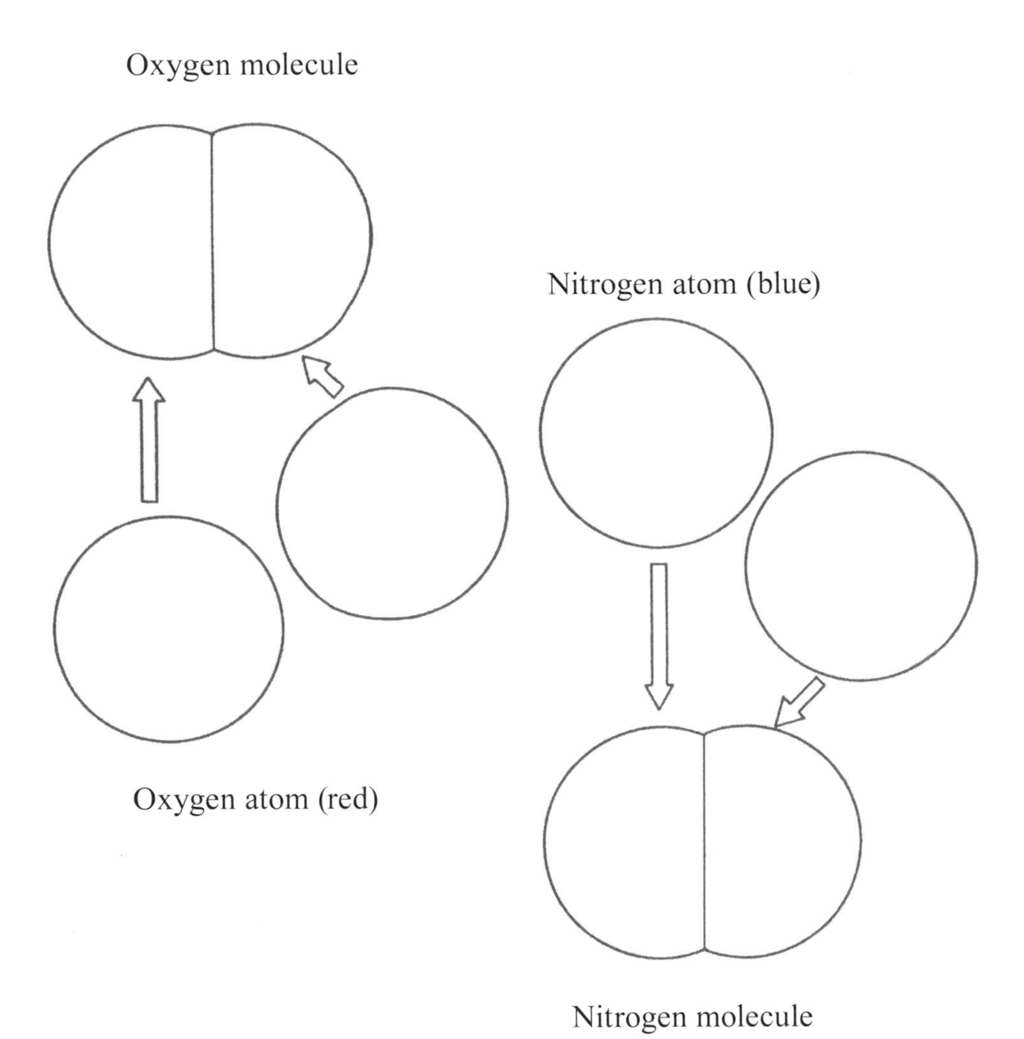

The size of the atom and the molecule is magnified 100 million times

Question 2

There are more nitrogen molecules than oxygen molecules in the air. The number of the nitrogen molecules is four times higher than the number of oxygen molecules.

Oxygen molecules

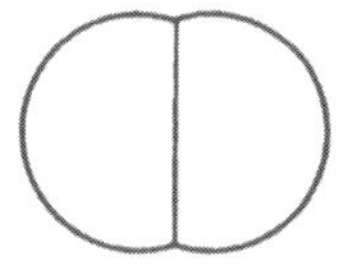

Nitrogen molecules

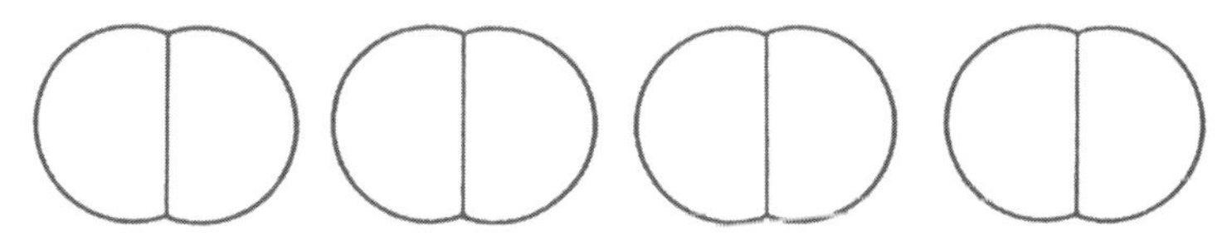

But, how many nitrogen and oxygen molecules are there in the air?

Imagine a model of the air made using these models of molecules magnified one hundred million times. Now imagine that we take a cubic section of the air that measures 0.000001 cm on each side and magnify it 100 million times. It will become a cube with each side measuring 100 cm (1 m).

How many nitrogen and oxygen molecule models should be put inside a '1 m × 1 m × 1 m' cube to make a model of the air?

What do you think?

a. About thirty thousand (30,000)
b. About three thousand (3,000)
c. About three hundred (300)
d. About thirty (30)
e. About three (3)

If the human head were magnified one hundred million times in size...

How big would an object be if it were magnified one hundred million times? To imagine this, how about thinking of the size of your head magnified one hundred million times?

The diameter of a person's head is about 15 cm. Therefore, if we were to enlarge it one hundred million times, it would be 1,500,000,000 cm. This is equal to 15,000 km, which is big!

The diameter of the Earth is about 13,000 km. Therefore, if a person's head were to be enlarged one hundred million times, it would grow bigger than the size of the Earth.

If we were to magnify your head and the molecules we are studying, each one hundred million times, your head would grow to a size greater than that of the Earth and an individual molecule would be about the size of a ping pong ball.

If the air were magnified one hundred million times…

When we magnify a cubic section of the air that measures 0.000001 cm on each side one hundred million times it will become a cube with each side measuring 100 cm = 1 m.

Make a cube this size with wooden frames.

This illustration of the model has been magnified ten million times.

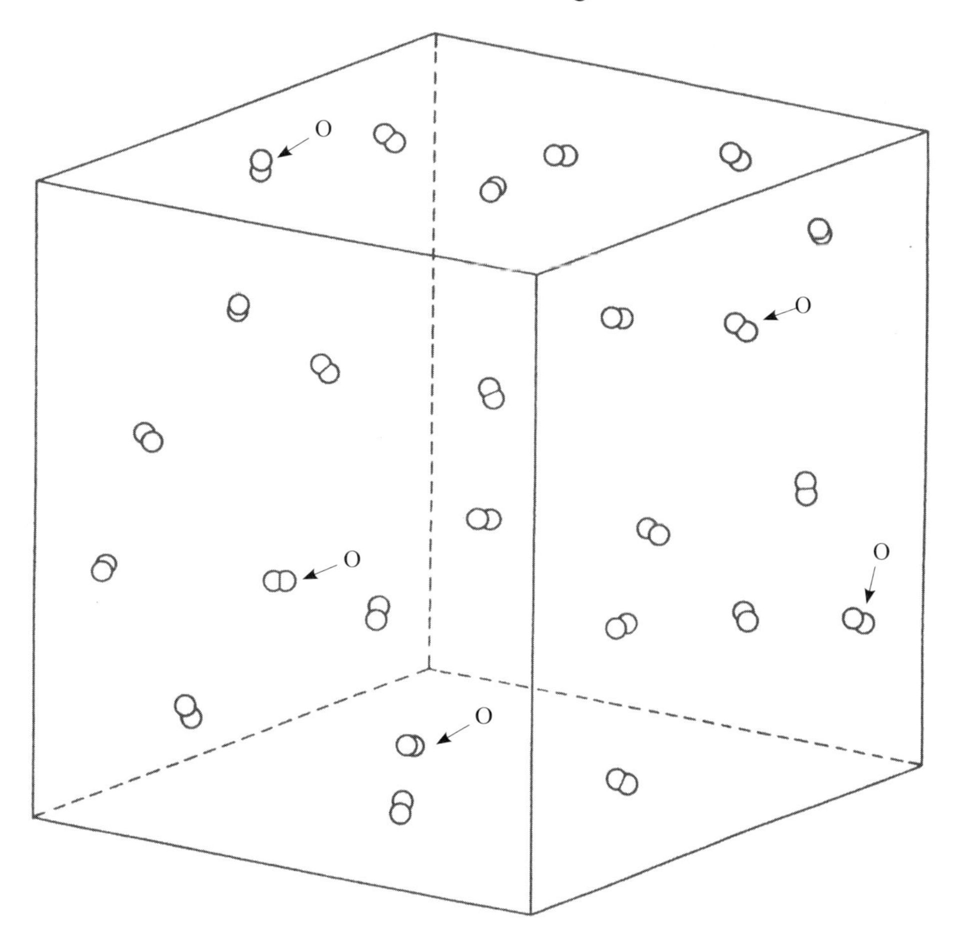

Colour the oxygen molecules (**O**) in red, and the nitrogen molecules in blue.

When scientists examine a cube measuring 0.000001 cm on each side, it seems to contain 25 air molecules. Therefore, in our magnified cubic model, which measures 1 metre on each side, there will also be 25 model air molecules. Those molecules have been magnified one hundred million times.

The number of nitrogen molecules in the air is nearly four times the number of oxygen molecules. We would, therefore, expect to find about five oxygen molecules among these 25 molecules.

Some of these molecules move at high speed but some are slow. On average, they move at about 400–500 m/s (1,400–1,800 km/h); sound moves through the air at 340 m/s. That means the speed of the molecules is 1.5 times faster than the speed of sound (Mach 1.5).

But there are a lot of molecules. Each molecule in the air collides with other molecules when advancing, on average, about 0.000014 cm in a straight line; on impact, it changes direction and collides with another molecule. It will collide with another molecule after advancing another 1,400 cm (14 m) in our model which has been magnified one hundred million times.

Of course, this is on average. There are likely to be molecules which collide at a shorter distance, and others that advance for longer without colliding.

Does the air contain any other molecules aside from nitrogen and oxygen molecules?

Look at the drawing of the air and search for other particles.

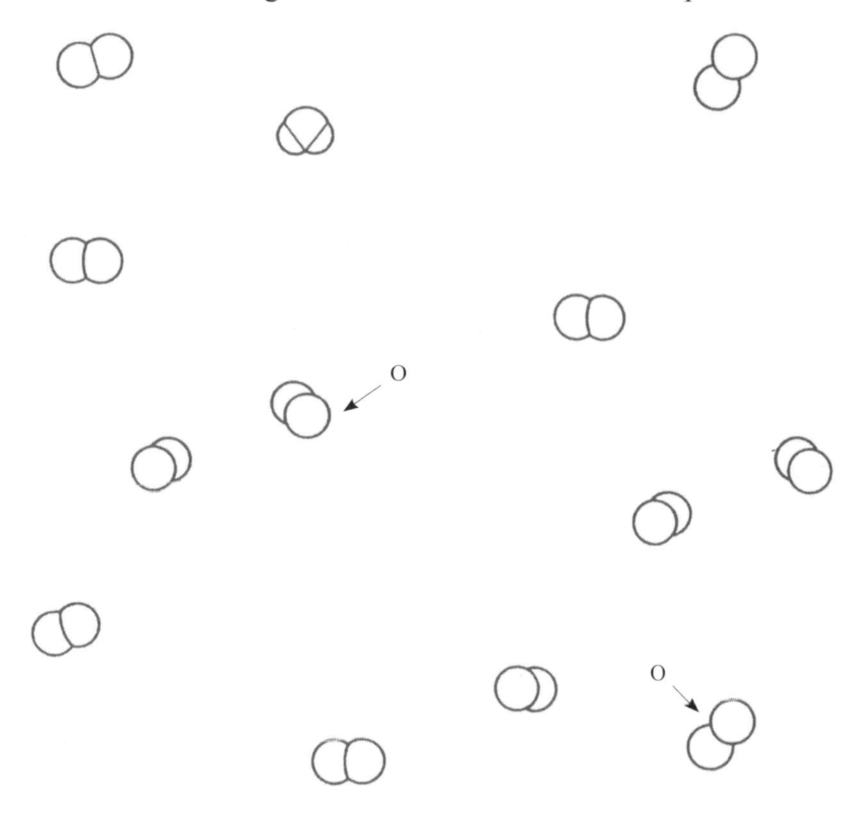

Colour the oxygen molecules red, and nitrogen blue.

Oh! Can you see another molecule? It is neither a nitrogen nor an oxygen molecule. It has a different shape.

The large atom in the centre is an oxygen atom. So, colour it in red.

The two small atoms that bond to either side of the red oxygen atom are called *hydrogen atoms*. Keep them white.

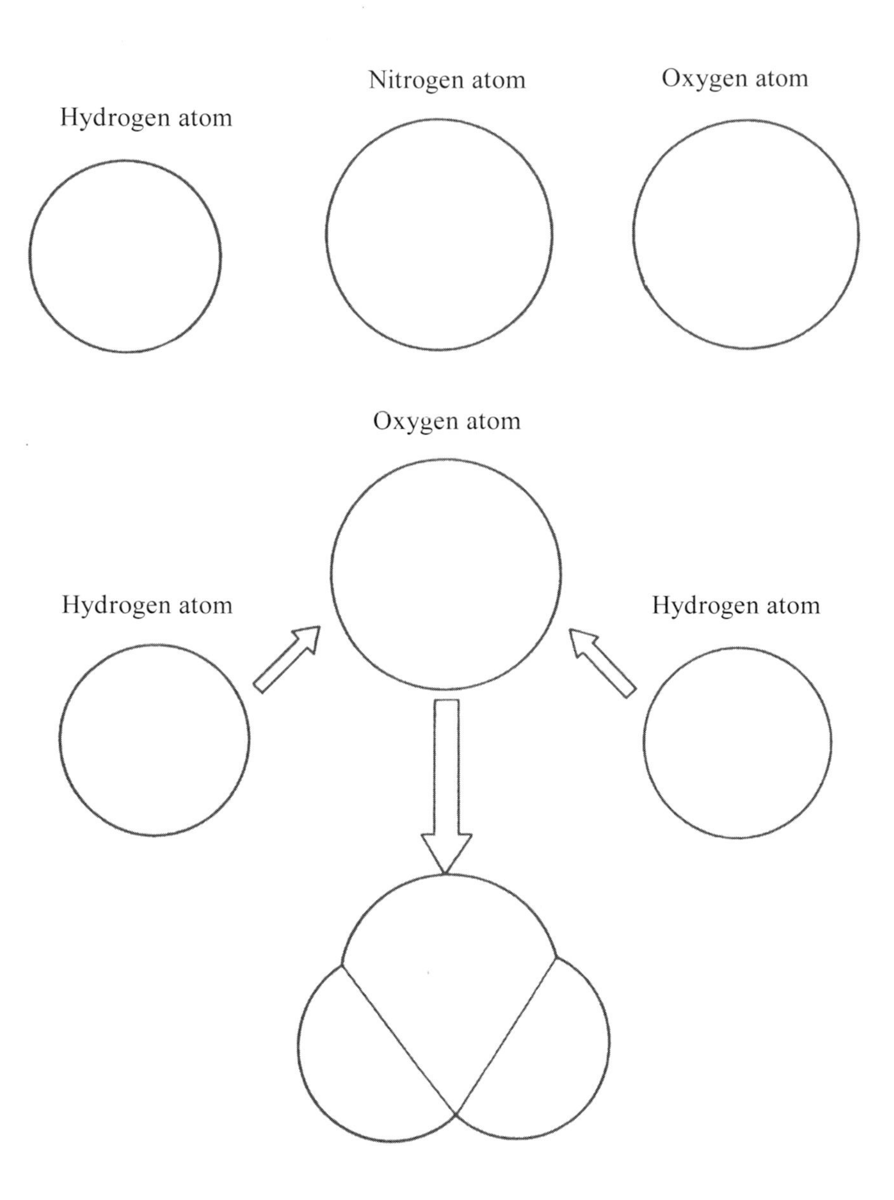

When two hydrogen atoms combine with one oxygen atom this is called a *water molecule.* Make a model of a water molecule.

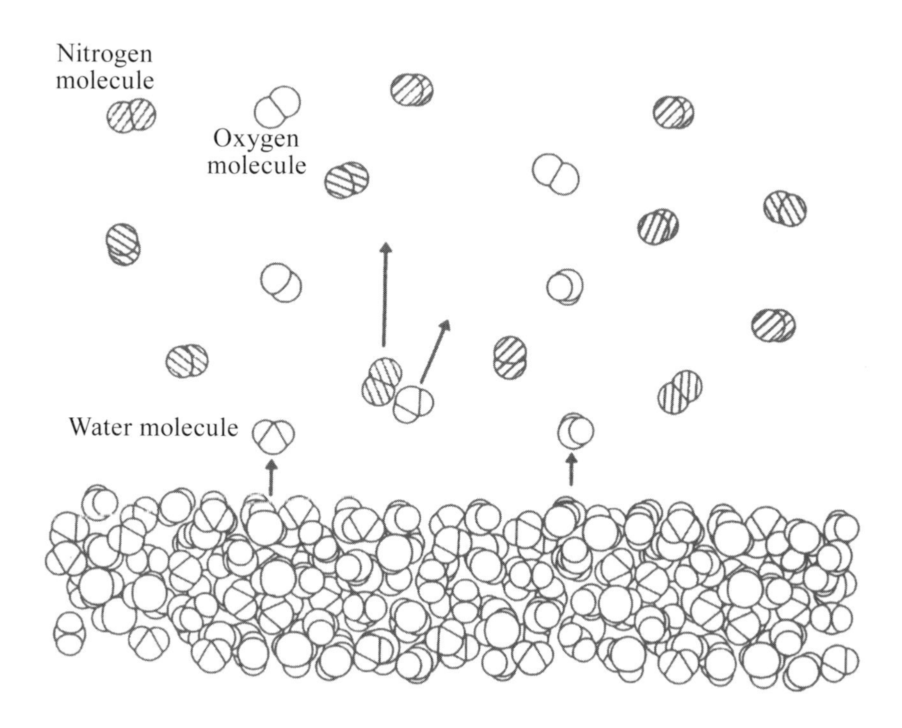

The water we see and use every day is made of many particles, much like the models you just made.

Why do you think that the water molecule is in the air?

If we leave water in a dish for some time, the level of the water might decrease without us noticing. This is because some of the water molecules from the dish escape into the air.

The water molecules also roam in the air as freely as the oxygen and nitrogen molecules.

The shape of the water molecule is strange.

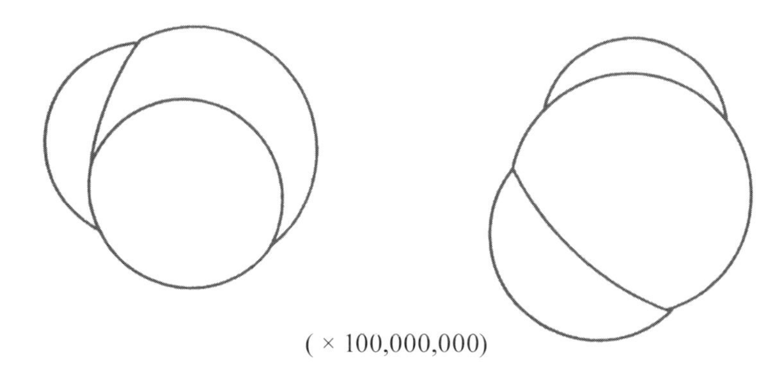

(× 100,000,000)

The one oxygen atom and two hydrogen atoms do not bond together in a straight line.

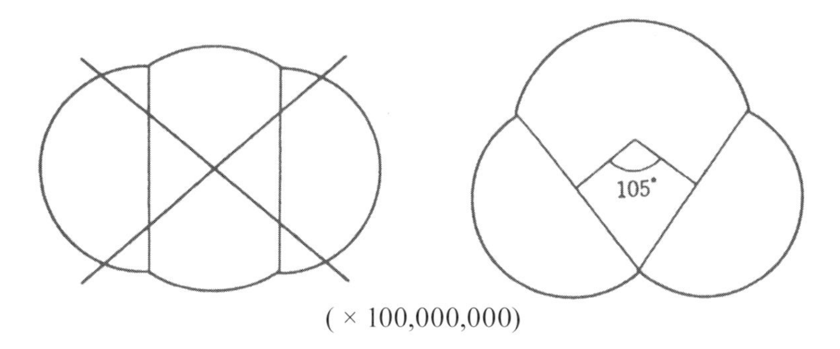

(× 100,000,000)

However, the way in which the two hydrogen atoms are combined with the oxygen atom is not random.

Two hydrogen atoms always maintain this same angle when they stick to the oxygen atom.

In the air, is there anything else apart from the nitrogen, oxygen and water molecules?

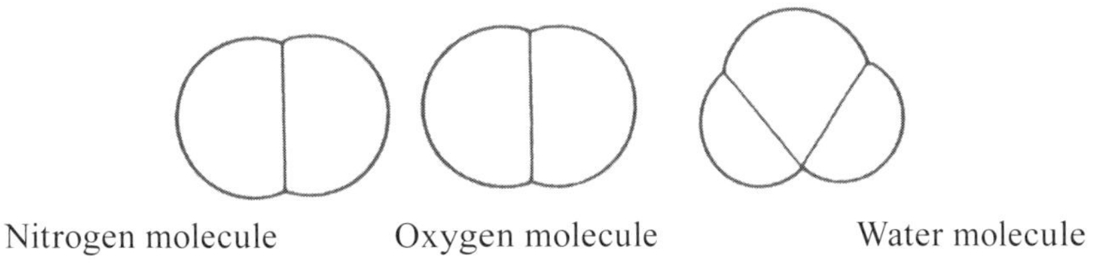

Look again at a drawing of the contents of the air. It contains a few other kinds of molecules.

Blank page

'A' in the drawing is *argon*. There is one argon molecule per every one hundred air molecules.

Draw a picture of an argon molecule next to the water molecule.

(× 100,000,000)

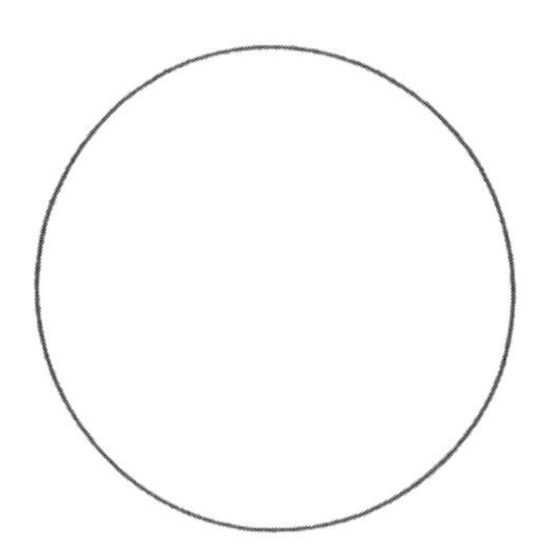

Argon molecule (atom)

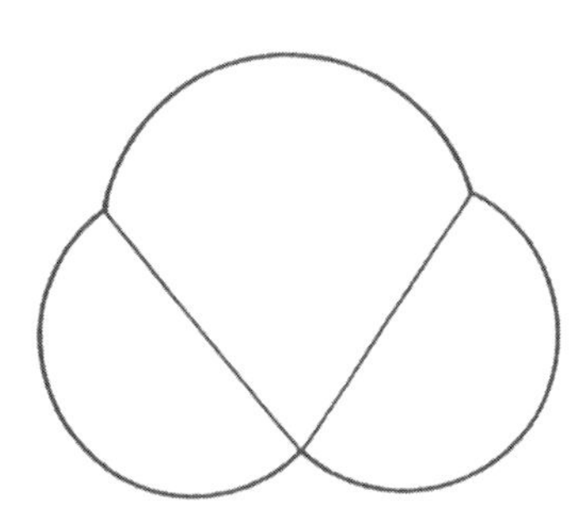

Water molecule

Colour the argon molecule purple.

The argon molecule is round no matter where you look at it from. Argon molecules also fly around in the air, but we find them in single atoms, unlike other molecules such as oxygen or nitrogen. Because of this you may use the term 'molecule' or 'atom' interchangeably when describing argon.

Two oxygen atoms or nitrogen atoms tend to bond together, but the argon atom does not do this. The argon atom also cannot combine with any other kind of atoms.

In the air there are still more atoms, like argon, that do not bond with others.

These are the *neon atom* and the *helium atom*.

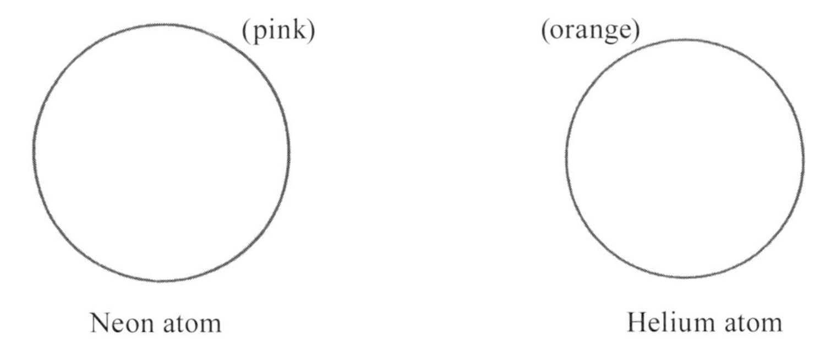

(× 100,000,000)

However, the air contains few neon and helium atoms. Only one or two are found in one hundred thousand air particles.

The molecule 'C' in page (*18*) is a molecule of *carbon dioxide*.

Two oxygen atoms combine with a *carbon atom*, and this is why it is called carbon dioxide ['di' means two].

A magnified carbon dioxide molecule is shown as below.

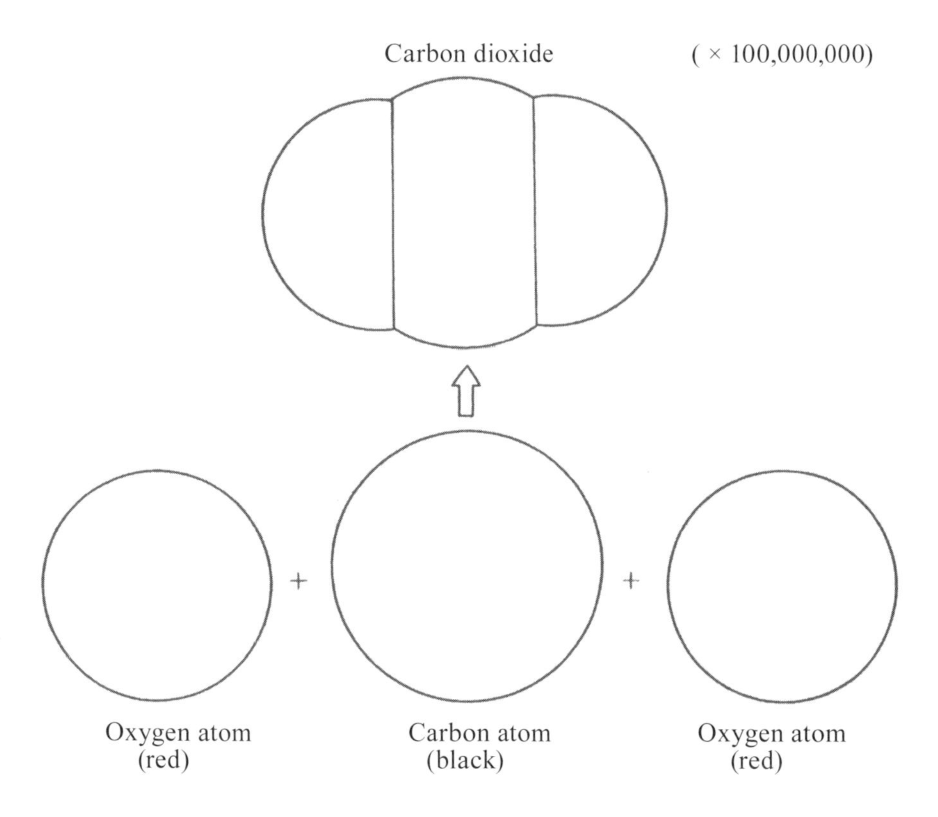

The three atoms that make carbon dioxide align in a straight row, and do not bend like the water molecule.

There are only four molecules of carbon dioxide in ten thousand air molecules.

A lot of carbon atoms are found in plants and animals. Charcoal is almost completely made of carbon atoms. When we burn wood and paper, the carbon atoms in the wood or the paper bond with the oxygen molecules in the air and make molecules of carbon dioxide. At this time the molecules of carbon dioxide move very fast and produce heat.

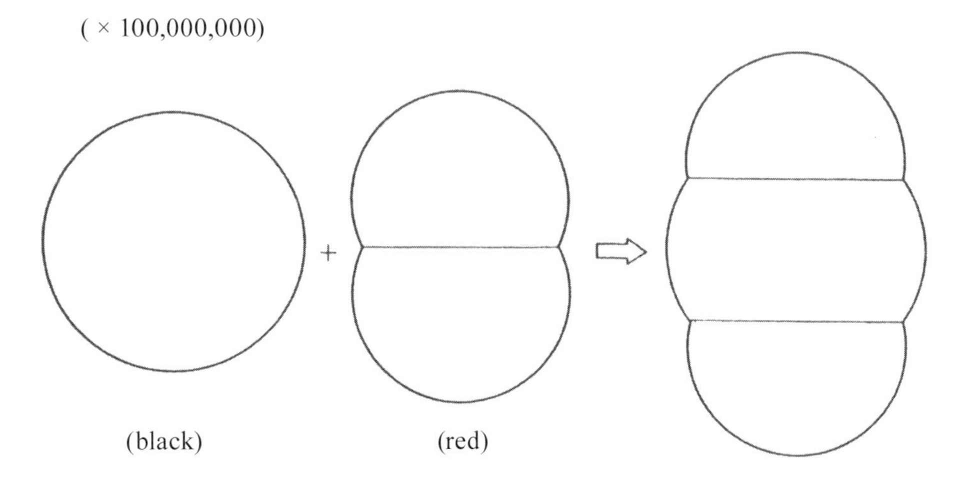

We breathe in order to take oxygen from the air into our bodies. As the carbon atoms in our bodies bond with the oxygen atoms from the air, the energy that allows us to move and warms our bodies is produced. When we breathe out, we release the carbon dioxide molecules that have been produced during this process.

This is what we find in every 100,000 air molecules.

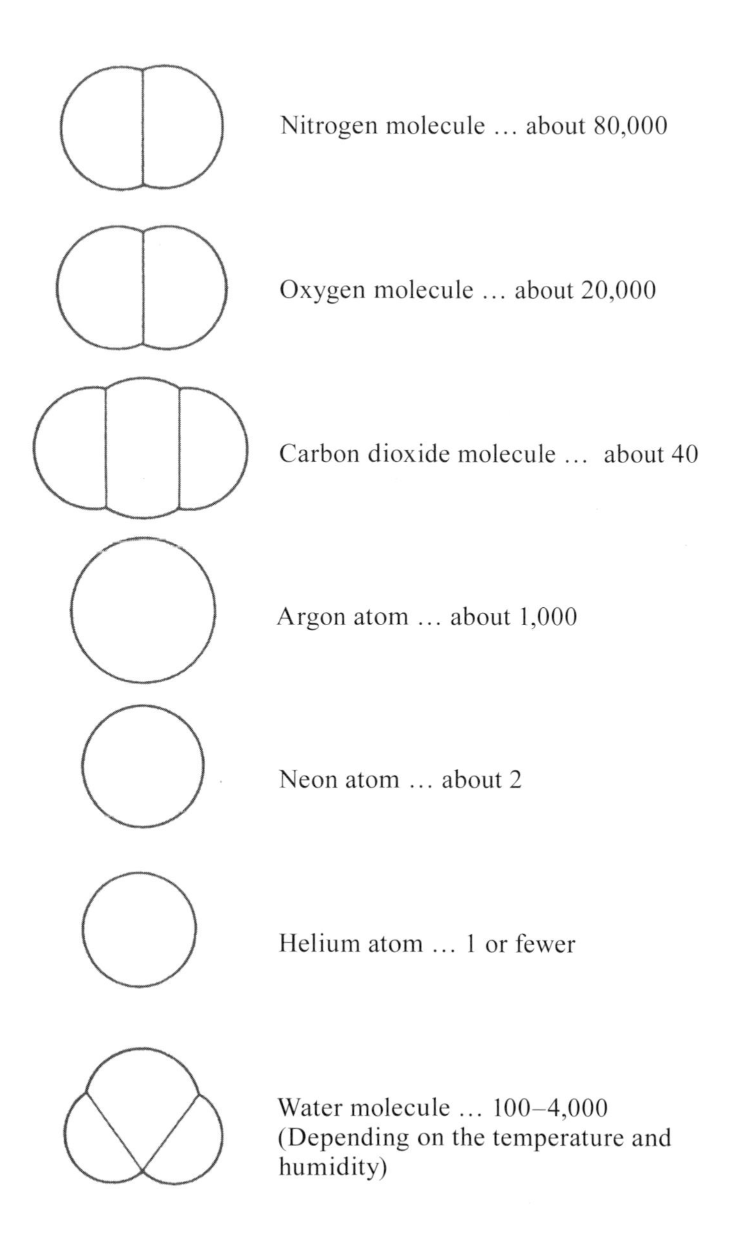

Blank page

In clean air, the molecules would appear as on the previous page. However, other molecules are present now in the air because human beings burn many kinds of things. Some of them are toxic and cause pollution. We will now look at the air in our cities which is dirty because of exhaust emissions from many cars and smoke from factories. What kind of molecules do you think that you could find in city air?

Have you ever heard the name of a gas that is harmful to plants, men, and other animals?

If you have heard of this gas, please tell everyone.

Blank page

Guess what this molecule is?

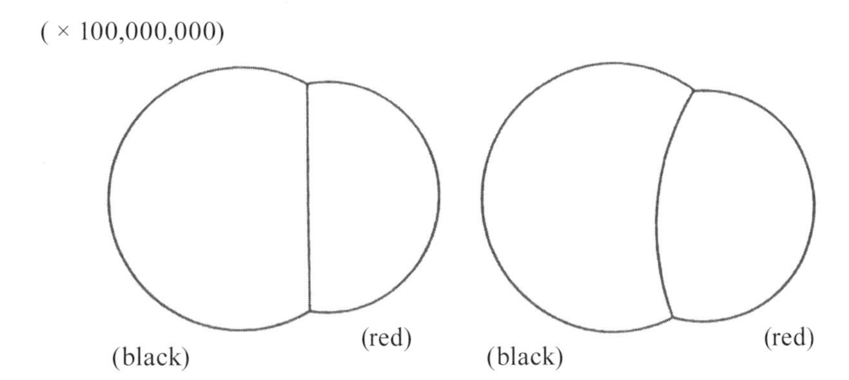

Do you recognise the black-coloured atom? It is the carbon atom.

What about the red-coloured atom? It is the oxygen atom.

This molecule is very much like carbon dioxide but has one fewer oxygen atoms. This molecule is called 'carbon monoxide' [Just like 'di' means two, 'mono' means one]. Do you think that will be easy to remember?

Carbon monoxide is released when fuels, such as gasoline and kerosene, are not burned completely. Carbon atoms are contained in gasoline and kerosene. When gasoline and kerosene are burned, two oxygen atoms will connect to each carbon atom and produce carbon dioxide. However, when there is insufficient oxygen, only one oxygen atom may bond with a carbon atom and carbon monoxide will be produced.

Carbon monoxide is found in the exhaust emissions of factories and vehicles. If kerosene or gas is burnt in a closed room, a considerable amount of carbon monoxide may be produced due to the lack of oxygen.

Carbon dioxide is not as toxic to the human body, animals, and plants as carbon monoxide, which is a seriously harmful molecule. After breathing large amounts of it even humans die quickly. When you burn kerosene or gas in your house, you should ensure that there is good ventilation.

Carbon monoxide is a flammable gas. If one more oxygen atom is bonded with it when it burns, it will become carbon dioxide. We can, therefore, remove carbon monoxide, which is harmful to the body, by burning it.

Recently, technology has been developed that burns carbon monoxide from exhaust gases or gases released from factories and changes it into carbon dioxide. Doing this has decreased the amount of carbon monoxide released into the air.

Next, what is this molecule?

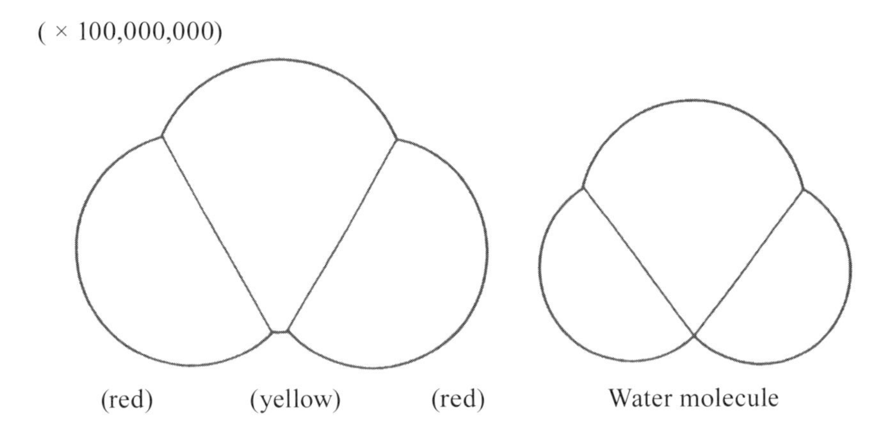

The yellow atom is a *sulphur atom*. Two oxygen atoms bond with an atom of sulphur, producing a shape similar to the water molecule. This molecule has two oxygen atoms and one sulphur atom, so we call it 'sulphur dioxide.' Occasionally this molecule is called 'molecule of sulphurous acid gas.' (Sulphurous acid is a solution of sulphur dioxide in water.) Oils, such as gasoline and kerosene that are extracted from underground, usually contain sulphur atoms. When we burn oil, sulphur atoms will react with oxygen and produce sulphur dioxides.

The sulphur dioxide molecule is very bad for human health. It is said that even one sulphur dioxide molecule in one million air molecules can harm the health of human beings. When sulphur dioxide dissolves into water, it becomes sulphurous acid or sulphuric acid. Both sulphurous acid and sulphuric acid are strong acids. If sulphur dioxide mixes with rain, this 'acid rain' will kill plants by causing them to starve.

Thus, sulphur dioxide is a harmful gas. Until recently, factory chimneys and cars had emitted large amounts of exhaust fumes that caused pollution. Sulphur dioxide cannot be produced without sulphur atoms. So, if we could use coal and oil which contain very few sulphur atoms, we would be able to decrease the amount of sulphur dioxide produced. Recently, technology that enables us to remove sulphur from oil has been developed, and now we can use 'gasoline and kerosene that contain little sulphur.' Moreover, another technology has been developed that captures sulphur dioxide and prevents it from being released into the air. As a result, sulphur dioxide from factories has been decreasing.

There is another harmful molecule which is emitted from factories and cars. It is the 'nitrogen dioxide molecule'.

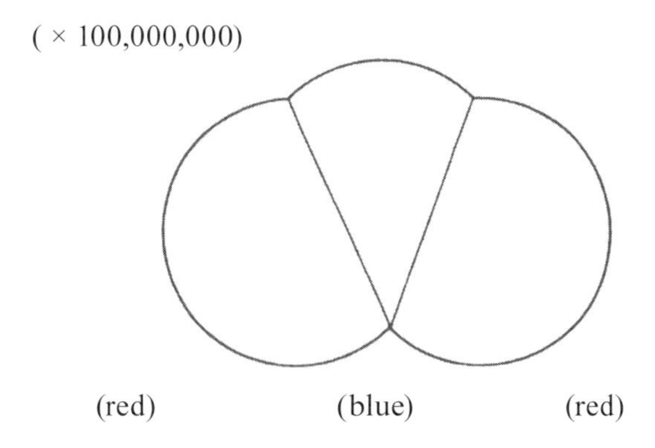

Nitrogen dioxide molecule

This molecule has two oxygen atoms and one nitrogen atom, so we call it 'nitrogen dioxide'.

There are a lot of nitrogen and oxygen molecules in the air. However, the two never bond naturally. But nitrogen dioxide can be produced in the presence of high temperatures; for example, in factories or in car engines.

Nitrogen dioxide is the main component of 'photochemical smog' pollution, which is harmful, including to human lungs.

Moreover, nitrogen dioxide also causes acid rain. When nitrogen dioxide dissolves in water, nitrous acid and nitric acid are produced. These are strong acids.

It is rather difficult to decrease the levels of nitrogen dioxide in the air. Unlike carbon monoxide, it cannot be changed into a harmless molecule by burning it. Moreover, the atom that is the critical component of nitrogen dioxide cannot be removed, unlike sulphur dioxide.

Air is used whenever factories or car engines burn fuel. Oxygen combines with nitrogen at a high temperature and it becomes nitrogen dioxide. At the same time, if a nitrogen atom bonds with an oxygen atom, this will produce a 'nitrogen monoxide molecule,' which is called 'Nitrogen oxide' or 'NOx (Noxious).'

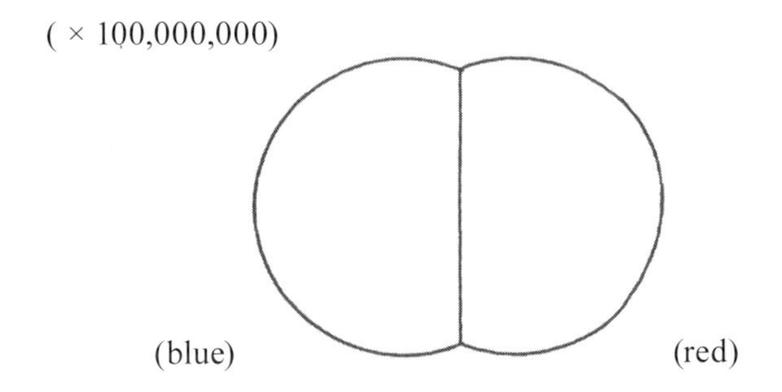

Nitrogen monoxide molecule

Question 3

Now please draw a picture of the air again. Imagine what it would look like 'if you could see the atoms and molecules in the air.' Please do not pay too much attention to the size of the molecules and the space between them. Use the space below or drawing paper.

Blank page

To contact the Association, please send emails to
mail@kasetu.co.jp

Translators
Kshitij Ghimire, Mariko Kobayashi

Translation advisers
Haruhiko Funahashi, Gianni Giosue,
Alexander Clemmens

Collaborators
Nobuo Takahashi, Tomoko Hasegawa,
Takashi Hasegawa, Hideyoshi Ueda,
Yoshinori Takeda

Editor
Mariko Kobayashi

Publisher
Kasetu-sha

Hypothesis–Experiment Classbook (*Jugyōsho*)

How Many Legs?

Association for Studies in Hypothesis–Experiment Class

Original Version 1971
English Version 2014

The aim of 'How Many Legs?'

The Classbook 'How Many Legs?' is used in elementary schools, from lower to higher grades.

The purpose of this Classbook is neither to make children memorise the number of animal limbs perfectly, nor to provide information about animal taxonomy: it is to encourage children to develop a broad understanding of the animal world from a simple point of view – the number of legs. It is also aimed at providing a broad and deep awareness about animals that cannot be obtained by simply visiting a zoo or looking at picture books.

By making predictions about unknown phenomena, we can observe nature more consciously. Such scientific experiences will enable us to make accurate predictions when we come across other unknown phenomena. In that sense, this Classbook does not focus on lessons for 'Learning about animals;' but, rather, it allows children to experience observing the animal world scientifically.

Do not scold children for their predictions regarding a problem by saying things like 'You don't even know that?' Never force students to simply learn answers. It is a good idea to prepare several types of illustrations and photo albums for verification of their predictions. It would be even better if there were real samples. In particular, butterfly images rarely have illustrations for the number of legs. Therefore, please prepare your own specimens if you can.

How Many Legs?

Question

In this picture we see some animals. Do you notice anything different about them? Is there something strange about their legs?

How about the animals on the next page?

On the next page, you will find the same animals with the correct number of legs.

Now we know that a lizard, a frog and alligator have four legs, and we know that a beetle has six legs.

We also know that a turtle has four legs and a duck has two legs.

A crane also has two legs, but sometimes it only stands on one leg when it folds one of its legs close to its body.

Part One: Ants

Question 1

Do you know how many legs an ant has?

Do you think you could draw an ant from memory?

Do you know how many segments the body has? Where are the ant's legs attached to its body?

Try to draw a big picture of an ant.

When you have finished, compare your drawings to a picture of a real ant.

Once you have looked at the picture try to draw an ant again.

(There is an accurate picture on the following page.)

Blank page

The story of the ant (1)

This is a Carpenter Ant. She is a worker in the nest. All of her legs grow from her middle. Compare your drawings to this picture.

How many legs does she have? Count them. Make sure you do not mistake her antennae, which grow from her head, with her legs.

The number of legs you counted is [ ].

Carpenter Ants consist of female worker ants, male drones, and a Queen ant. Male drones and the Queen ant both have wings, but the Queen loses her wings before she starts to lay her eggs.

Problem 1

The ants that live around our school come in many different shapes and colours.

Before going out to take a look together, share your ideas with your classmates.

(1) How many different kinds of ants do you think we will find around our school?

Expectation

About [ ] kinds.

(2) Do you think that they will all have 6 legs?

Expectation

a. Yes, they will all have 6 legs.
b. Some ants might have [ ] legs.

Results

Kinds of ants [ ]

Number of legs [ ]

Blank page

Problem 2

(1) How many kinds of ants are there in your country?

Expectation

a. About 5 kinds
b. About 10 kinds
c. About 50 kinds
d. Many more kinds

(2) Do ants all over the world all have 6 legs?

Expectation

a. They all have 6 legs.
b. There might be some ants which have a different number of legs.

Discuss with your classmates.

Next, look at some books with pictures of ants.

Blank page

The story of the ant (2)

In Japan there are more than three hundred different kinds of ants. They come in many different sizes, colours, and shapes. Some scholars believe that there are more than 20,000 different kinds of ants all across the world. However, every ant has six legs and their bodies all consist of three parts. If we start from the front of the ant we refer to the parts as the 'head,' the 'thorax,' and the 'abdomen.'

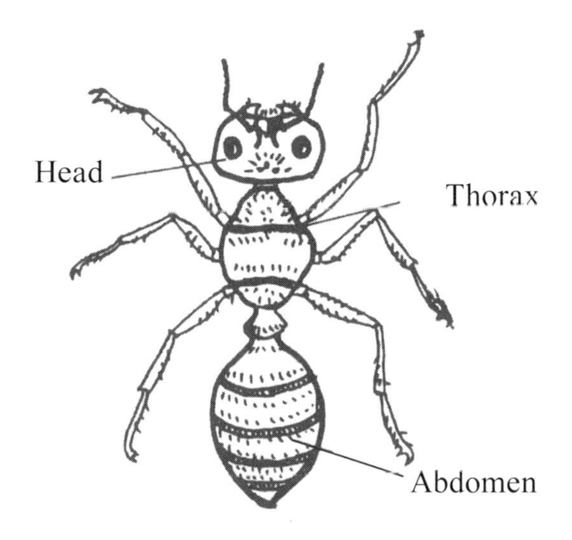

All six of the ant's legs grow from its thorax. You might think that there would be some ants that have as few as four, or maybe as many as eight, legs. However, every ant, no matter how big or how small, has exactly six legs. If you find an ant that does not have six legs, it has probably suffered an injury.

Blank page

Part Two: A variety of bugs

Problem 1

How many legs do you think the Monarch Butterfly and the Cabbage White Butterfly have?

Expectation

[ ] legs

If we can catch a butterfly, we can count the number of legs it has. If you do not have a butterfly handy, on the following page there are some pictures of the Monarch Butterfly and the Cabbage White Butterfly that we can examine.

(If you catch a butterfly, do not forget to let it go!)

Blank page

How many legs does the Monarch Butterfly have?

☐ legs

How many legs does the Cabbage White Butterfly have?

☐ legs

Problem 2

How many legs do a dragonfly and a cicada have?

Expectation

The dragonfly has ☐ legs.

The cicada has ☐ legs.

If we can catch one, then we can count its legs.

If there are not any to catch, look at the pictures on the next page to find out.

How many legs does a dragonfly have?

☐ legs

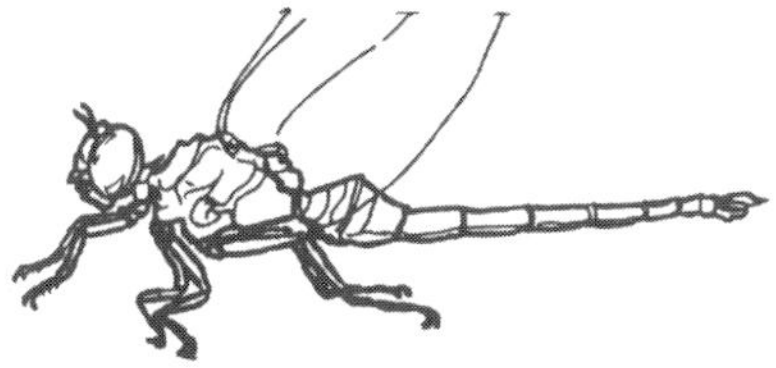

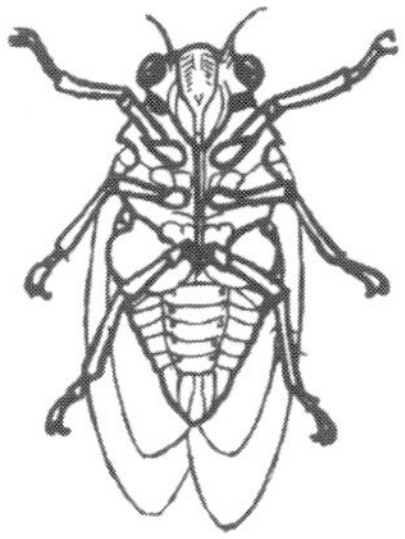

How many legs does a cicada have?

☐ legs

Problem 3

How many legs do a cricket, a beetle, and a housefly have?

A cricket has ☐ legs.

A beetle has ☐ legs.

A fly has ☐ legs.

Check your guesses by catching the bugs or looking at some pictures in a book.

Problem 4

Do you know any other bugs that we have not looked at yet?

How many legs do they have?

Guess and then check the number of legs.

Bug's name	Number of legs	
	Your guess	Actual number
Bee		
Spider		
Mosquito		

Bugs' legs

It is common for many insects to have six legs. As we have learned, ants, butterflies, dragonflies, crickets, beetles and flies all have six legs.

In the case of the praying mantis, it may appear to have only four legs, but if you look very carefully you will find that its two pincers, called raptorial forelegs, are also legs, so it too has six legs.

At first glance, it appears that the water strider has only four legs as well. However, if you look carefully you will notice that it also has six legs.

The spider is different from other bugs. It actually has eight legs.

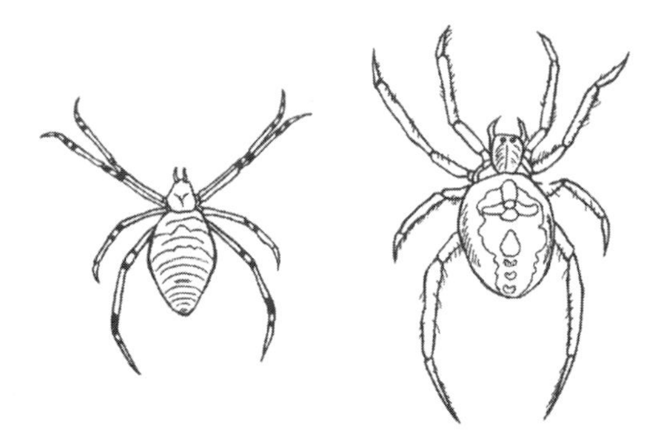

When examining a spider please be careful not to confuse its antennae and its front legs. A spider belongs to a different class of bugs. Scientists classify bugs which have six legs as *insects*.

A ladybug has six legs, and so we know that it is an insect too.

A grasshopper has _____ legs, so it is __________ .

Is a cockroach an insect? ☐

Is a firefly an insect? ☐

Is a frog an insect? ☐

We have learned that most bugs, such as insects, have six legs, but some bugs, like the spider, have eight legs. A scorpion and a tick have eight legs, so they are not insects.

Problem 5

Are there any bugs that have only four or two legs?

Expectation

(1) a. I think there are bugs with four legs.
    b. I do not think there are bugs with four legs.

(2) a. I think there are bugs with two legs.
    b. I do not think there are bugs with two legs.

After making your predictions, look at some pictures to see if you were right.

Then continue on to the next page.

Blank page

Are there bugs with four legs or two legs?

Do you think you might be able to find a bug that has only four legs, or maybe only two legs? Do you think it will be difficult to find them?

You think you already found one? Really?!?!

I think you might have made a mistake. Take a closer look at that bug. Actually, there are no bugs with four or two legs. There are hundreds of thousands of different kinds of bugs, so it might not seem strange that there would be bugs with four or two legs. But, in fact, we know of no bugs like this.

Really? You might think, 'A frog has four legs, doesn't it?' You would be right, it does. But we already know that a frog is not a bug.

We know this because a frog, unlike a bug, has a spine that is contained inside its body. If an animal has a spine, it is not a bug.

A lizard has a spine, too.

Frog skeleton

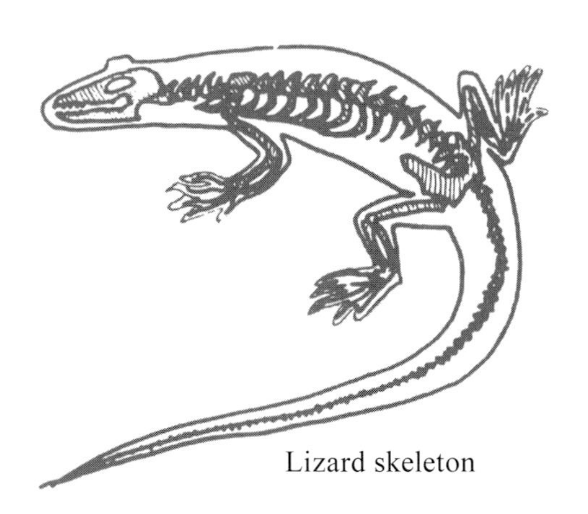

Lizard skeleton

Insects do not have a spine. If you do not believe this, look at the pictures below.

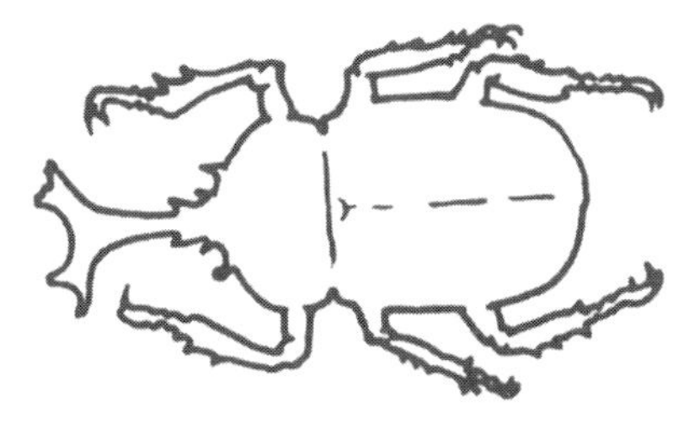

The beetle has no spine.

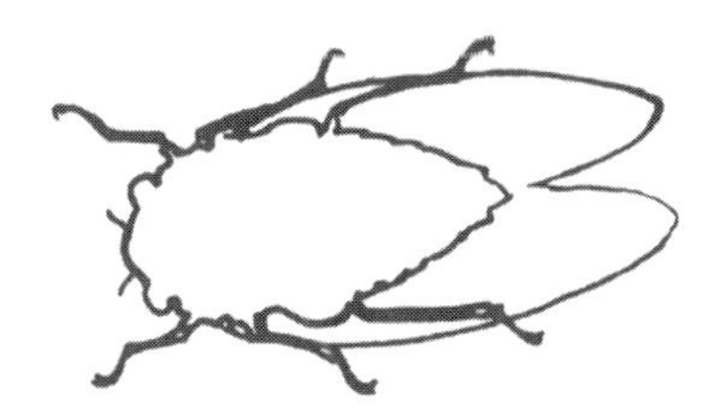

The cicada does not have a spine either.

There are many bugs which have eight or ten legs, but there are no bugs with four or two legs.

Blank page

Part Three: The legs of animals that have a spine

Large land-dwelling animals have a spine. It is the spine that allows them to become so large. The spine is able to support the weight of their bodies, so they may grow much larger than a bug and not collapse.

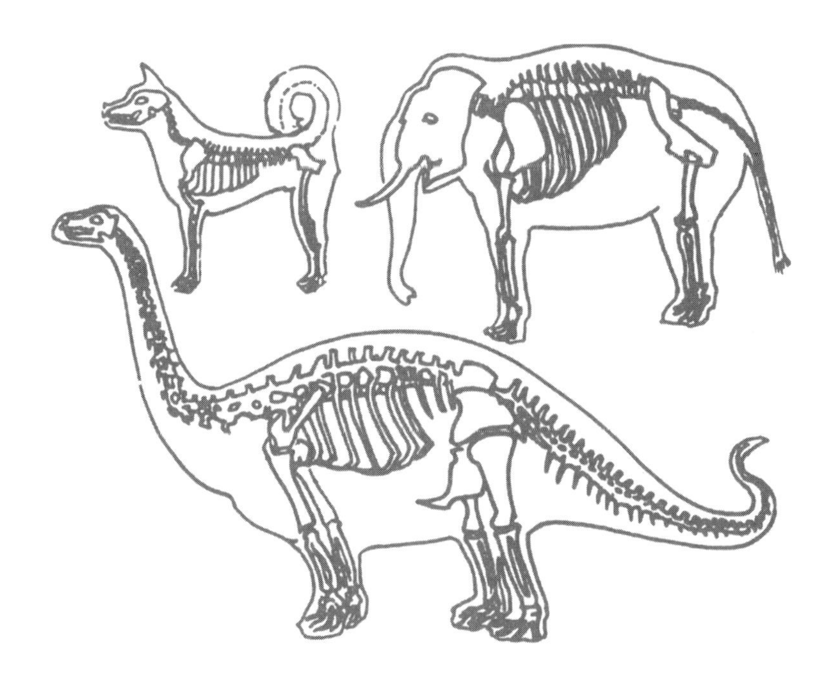

Question 1

Many animals that have a spine also have four legs.

How many animals with four legs can you name? Write down as many as you can.

Problem 1

Are there animals that have a spine and only two legs?

Expectation

a. No
b. Yes, for example:

Discuss your guesses and then continue on to the next page.

Animals with two legs

One example of animals which have two legs and a spine are birds. Instead of a second pair of legs, birds have wings. We classify birds as any animal with two legs, a spine, and feathers covering its body.

On this and the next page, we have pictures of penguins and ostriches. Although they are unable to fly, should we still consider them to be birds?

They do have two legs and their bodies are covered by feathers. Also, both penguins and ostriches, in common with other bird species, have a beak and no teeth. Because of these facts, we do consider them members of the bird family.

Problem 2

Do you think there are any animals that have a spine but no legs?

Expectation

a. No
b. Yes, for example: [ ]

Discuss your guesses and then continue on to the next page. You can confirm if you were right by using pictures from books.

The spine and legs

We have two examples of animals that have a spine but no legs – fish and snakes.

Because fish live in water they do not need legs to move; they can swim.

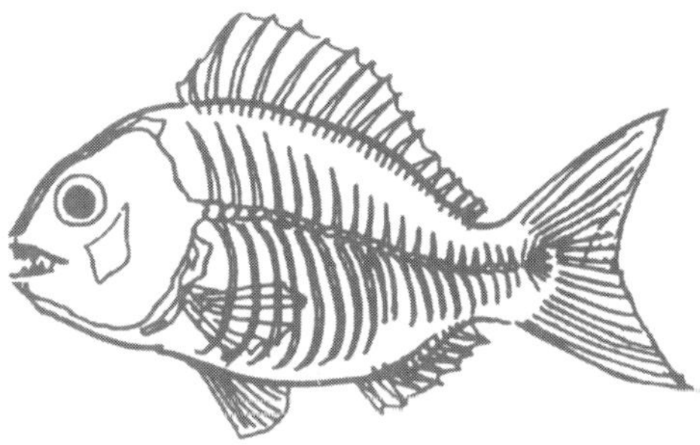

They use their fins to swim. Four of the fins that fish have, the two pectoral fins and the two pelvic fins, are structurally very similar to legs.

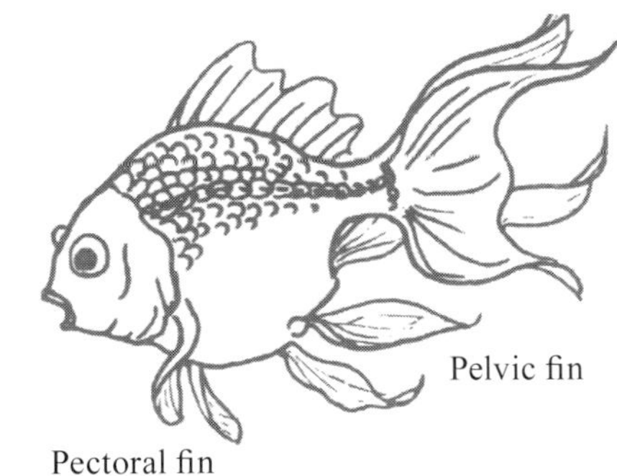

Within the fish family there are several examples of fish that can leave the water and crawl across land. On these fish, the pectoral fins act like legs, giving the fish the ability to move on land. Examples of these fish are the Mudskipper and the Northern Snakehead Fish. We can get an idea of how these fish move on land by looking at pictures of them.

There is another fish, known as the Coelacanth (pronounced See-la-canth) whose pectoral fins are very thick and appear more like legs.

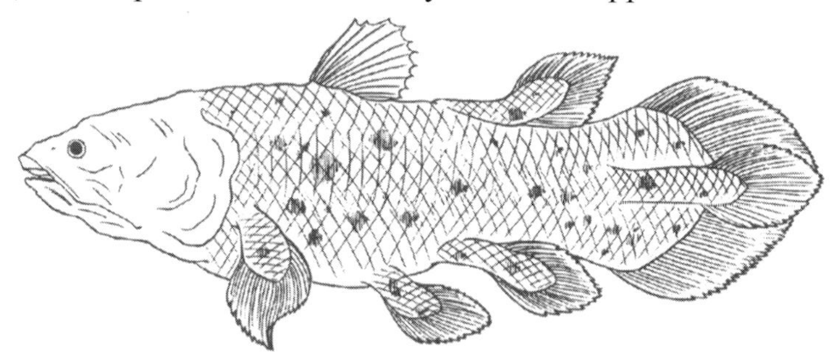

Snakes also do not have legs. It may appear that snakes do not have a spine, but they actually do.

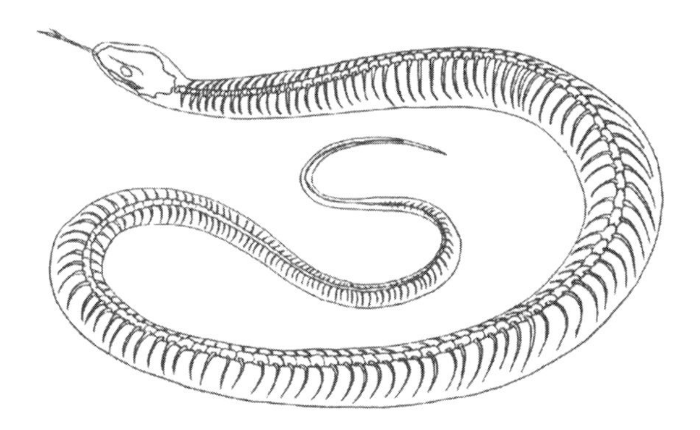

Earthworms appear very similar in shape to snakes, however, as they lack a spine, they are not considered to be related to snakes.

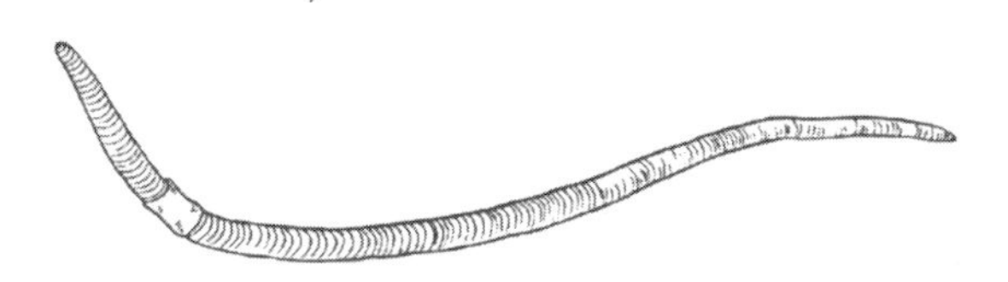

As a rule, animals which have a spine also have either four, two, or no legs.

'What's that?!' you say, 'Crabs and octopuses have eight legs?!'

This is correct, but the octopus and the crab do not have a spine.

Within the squid and the crab there is something similar to a spine, which has the appearance of scotch tape, but it is not actually a spine.

Scientists classify the spine as a construct of small linked bones.

Animals with many legs

Among those animals that lack a spine there are several kinds that have many, many legs; the centipede and millipede are well known examples.

Count the legs of the centipede in the picture below.

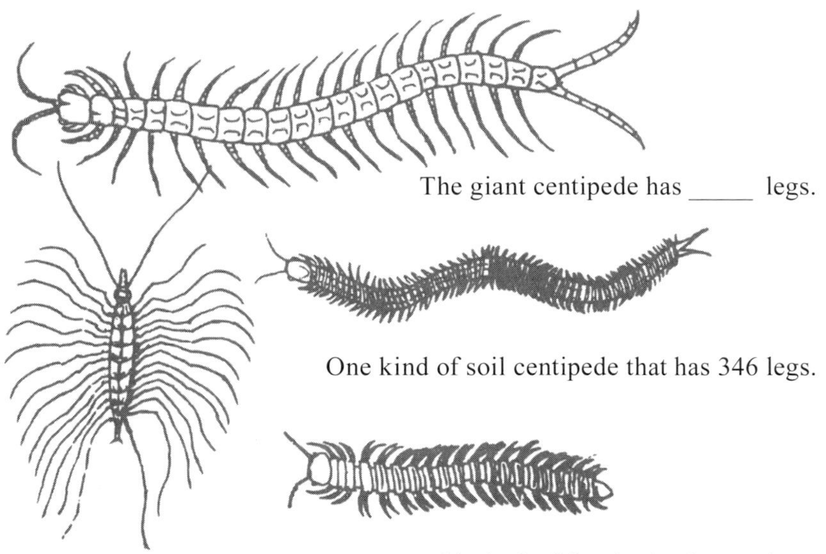

The giant centipede has _____ legs.

One kind of soil centipede that has 346 legs.

A kind of millipede that has 70 legs.

The house centipede has 30 legs.

The giant centipede has 42 legs, but we know that some kinds of centipede have 346 legs.

From now on, whenever you see an animal, look closely at its legs and spine.

The end

To contact the Association, please send emails to mail@kasetu.co.jp

Translator
Mariko Kobayashi

Translation advisers
Gianni Giosue, Robert Cadwallader
Haruhiko Funahashi

Colaborator
Hiroaki Arai

Editor
Mariko Kobayashi

Publisher
Kasetu-sha, http://www.kasetu.co.jp/